T0201155

Distributed Cooperative Control

Distributed Cooperative Control

Emerging Applications

Yi Guo

This edition first published 2017
© 2017 John Wiley & Sons, Inc.

The right of Yi Guo to be identified as the author of this work has been asserted in accordance with law.

Registered Office
John Wiley & Sons, Inc., 111 River Street, Hoboken, NJ 07030, USA

Editorial Office
111 River Street, Hoboken, NJ 07030, USA

For details of our global editorial offices, customer services, and more information about Wiley products visit us at www.wiley.com.

Wiley also publishes its books in a variety of electronic formats and by print-on-demand. Some content that appears in standard print versions of this book may not be available in other formats.

Library of Congress Cataloging-in-Publication Data

Hardback: 9781119216094

Cover image: Mina De La O/Gettyimages

Set in 10/12pt Warnock by SPi Global, Pondicherry, India

Printed in the United States of America

10 9 8 7 6 5 4 3 2 1

To My Parents

Kuisheng Guo and Zhenfu Liu

Contents

Preface

Networked systems are most familiar in the form of data networks connecting computers (*e.g.*, Internet) or networks for voice communications. However, in the broader sense, networks provide the infrastructure for spatially distributed components of a system, or distributed systems, to intercommunicate, enabling powerful capabilities for cooperative and collective behaviors for the overall system. Such collective behaviors are seen in man-made systems (*e.g.*, large power systems and transportation networks) as well as in naturally occurring systems (*e.g.*, flocks of birds, clusters of biological organisms, and gene networks). The remarkable advances in networking technologies (both wired and wireless networks) are enabling a wide range of new, man-made systems based on sophisticated and low cost distributed components cooperating across low cost and high-speed data networks. Also, the availability of new tools and the development of new techniques in handling multi-scale systems (from micro aerial vehicles to nanoparticles) give unprecedented opportunity for breakthroughs in controlling realworld networked systems.

Cooperative control of multi-agent systems has experienced rapid development during the past few decades. Distributed consensus algorithms received much research attention, and cooperative control of multi-vehicle systems have been extensively studied extending information consensus to coordinated motions of autonomous vehicles. Due to domain specifications and system constraints, applications found in realworld systems pose new challenges to cooperative control. The richness of realworld problems and diverse application domains revitalize the field and stimulate new research directions.

This book presents applications of distributed cooperative control to realworld engineering and physics systems addressing emerging needs for high efficiency distributed control systems. After introducing backgrounds and reviewing fundamental distributed consensus algorithms, technical contents of the book are divided into three parts. The first part deals with networked communication systems, and consists of three chapters, which present distributed consensus for quantized communication, cooperative spectrum sensing and distributed radio environment mapping for cognitive radio networks. The second part of the book presents cooperative control of multi-robotic systems, and includes two chapters to discuss the source seeking and plume tracking problems by distributed cooperating robots. In the third part of the book, cooperative control of multi-agent physics systems are addressed, which includes two chapters on friction control of coupled nanoparticles and synchronization of coupled laser arrays. Despite variations in system scales (from macroscale robots to nanoscale particles) and differences of application domains (from electrical to physics systems), the fundamental control issue to maintain coordination or synchronization of system components is universal, which is captured either by the fundamental concept of consensus/agreement, or through information diffusion over a network to obtain distributed estimation of global information. The graph matrix tools are also instrumental to stability and convergence analysis throughout the book.

The results reported in the book summarize the research effort on cooperative control applications of my research group at Stevens Institute of Technology during the past few years. The graduate students, Shuai Li, Wenlin Zhang, Ruofan Kong, and the postdoctoral researcher, Zheng Wang, directly contributed to the research reported in the book. The book uses materials from the Ph.D. dissertations of Shuai Li and Wenlin Zhang. The reported work was generated through research collaborations with my colleagues at Stevens and other institutions including Hongbin Li (Stevens), Yingying Chen (Stevens), Zhihua Qu (University of Central Florida), Yehuda Braiman (Oak Ridge National Laboratory), Brian Bingham (University of Hawaii and Naval Postgraduate School), Joseph Mitola III (Federated Wireless), and Zhenyu Zhang (University of Science and Technology of China). I would like to thank my students for their hard work, and thank my collaborators for introducing me to control problems originated from other fields.

Acknowledgment is given to National Science Foundation for providing funding to support this work during the past few years under Grants CMMI-0825613, EFRI-1024660, IIS-1218155, CNS-1318748, and IIS-1527016. Finally, I want to thank my parents (Kuisheng Guo and Zhenfu Liu), my husband (Weimin), and my kids (Raymond and Victoria) for their love and support.

April 2016

Yi Guo
Hoboken, New Jersey

About the Companion Website

This book is accompanied by a companion website:

www.wiley.com/go/Guo/DistributedCooperativeControl

The website includes:

- HD version of all the illustrations from the book

1

Introduction

1.1 Motivation and Challenges

The current book investigates emerging applications of multiagent cooperative control. It is motivated by the ubiquity of networked systems and the need to control their behaviors for real-world applications. We first review collective behaviors and then introduce major technical challenges in cooperative control.

1.1.1 From Collective Behaviors to Cooperative Control

Collective behaviors are observed in natural systems. Groups of ants create colony architectures that no single ant intends. Populations of neurons create structured thought, permanent memories, and adaptive responses that no neuron can comprehend by itself. In the study of collective behaviors, usually some types of agent-based-models are expressed with mathematical and computational formalisms, and the descriptive model is capable of quantitative and objective predictions of the system under consideration. The descriptive equations of fish schools and other animal aggregations were proposed in Ref. [1] in the 1950s, and it is more than three decades later that renewed mainstream attention has been received in a range of fields—including computer graphics, physics, robotics, and controls. A distributed behavior model, which is based on the individual agent's motion, is built by Reynolds [2] and computer simulations are done therein for flock-like group motion. Individual-based models and simulation of collective behaviors are also addressed in Ref. [3] with discussions of collective effects of group characteristics. Simulated robots are used in Ref. [4] to simulate collective behaviors where different types of group motions are displayed. While the aforementioned work is mainly on descriptive models and simulated behaviors, controlling the movement of a group using simulated robots with dynamic motion is addressed in Ref. [5]. Collective behaviors such as seen in herds of animals and biological aggregations are also referred to as swarming in the literature. Models of swarming are discussed in Refs. [6, 7], where attraction–repulsion interactions

Distributed Cooperative Control: Emerging Applications, First Edition. Yi Guo.
© 2017 John Wiley & Sons, Inc. Published 2017 by John Wiley & Sons, Inc.
Companion website: www.wiley.com/go/Guo/DistributedCooperativeControl

are included in the system's dynamics. Stability analysis of swarms is given in Refs. [8, 9] based on certain artificial interaction forces. The research has progressed rapidly in recent years from modeling and simulation of specific examples toward a more fundamental explanation applicable to a wide range of systems with collective behaviors.

In physics, the phenomenon of collective synchronization, in which coupled oscillators lock to a common frequency, was studied in the early work [10, 11]. In the 1970s, Kuramoto proposed a tractable model (referred to as the Kuramoto model) for oscillator synchronization [12, 13]. A related problem, the collective motion and phase transition of particle systems, is considered from the perspective of analogies to biologically motivated interactions in Refs. [14, 15] where simulated behaviors are presented. The studied models are capable of explaining certain observed behaviors in biological systems, including collective motion (rotation and flocking) of bacteria, networks of pacemaker cells in the heart, circadian pacemaker cells in the nucleus of the brain, metabolic synchrony in yeast cell suspensions, and physical systems such as arrays of lasers and microwave oscillators. Despite 40 years having elapsed since Kuramoto proposed his important model, there remain important theoretical aspects of the collective motion that are not yet understood; see Ref. [16] for a review on the topic.

More recently, the phase transition behavior described by Vicsek and coauthors [15] was revisited and theoretically explained by Jadbabaie et al. [17]. Their work is significant since it provides a graph theory–based framework to analyze a group of networking systems. Since then, coordination of mobile agents has received considerable attention. The consensus problem, which considers the agreement upon certain quantities of interest, was posed and studied by Olfati-Saber and Murray [18]. Here, the network topology was explicitly configured and the relationship between this topology and the system convergence was addressed using graph theory–based methods. The problem is further studied in Refs. [19–22], and necessary and sufficient conditions are given for a networked system to achieve consensus with the switching topology (see a survey [19]). Subsequent studies extended the principles of cooperative control to applications related to vehicle systems, for example, in Refs. [23–29]. Since then, cooperative control has gone through periods of rapid development [30–33].

1.1.2 Challenges

Despite rapid development, the field of cooperative control is far from mature. Major technical challenges arise from system dynamics and network complexity, which include the following:

- *Nonlinear agent dynamics*: Most agent systems are nonlinear dynamic systems. For example, cooperating robot vehicles, such as ground, aerial, and underwater vehicles, are *nonlinear* dynamic systems: the states of the system vary in time in complicated ways. Most existing cooperative control based on graph theory methods assumes single integrator or simple linear dynamics, which is not adequate for real-world applications where the performance of the designed control system can deviate greatly from the performance suggested by these simplified system models. Design of cooperative control for nonlinear systems is not a trivial task. There is no general framework available for nonlinear cooperative control.

- *Nonlinear agent interactions*: In many natural systems, the adhesive and repulsive forces among agents are nonlinear. For example, the repulsive force between two agents may need to become very large (and approach infinity) when they are very close in order to avoid collisions. Similarly, when the distance between them is greater than a threshold, the repulsive force may either become very small or vanish. Most existing cooperative control framework addresses linear agent interactions, while many real-world multiagent systems, such as nanoscale particle systems, have complicated nonlinear interactions, for example, Morse-type interactions. New methodologies are called upon to solve cooperative control problems to support systems with more general (nonlinear) agent interactions.

- *Robustness*: Due to uncertainties in agent dynamics, communication links, and operating environments, robustness has to be considered for a successful system design. For example, uncertainties in the communication links of cognitive radio networks (CRNs) include time delay of information exchange and time-varying and/or switching of the network connectivity. Such uncertainties can lead to unexpected or perhaps unstable behaviors. Robustness consideration has been discussed for the basic consensus problem in existing work, but general robust cooperative control for complicated real-world systems has not been adequately addressed. The ultimate goal is that, under a well-designed control scheme, the closed-loop networked system will be tolerant of and robust to network and environment disturbances.

- *Diversity of real-world problems and application domains*: Networked systems are becoming increasingly ubiquitous. Depending on the domain of applications, the control objectives and constraints are inherently different. For example, control of nanoscale particle systems has strict confinement constraints, the system is not readily accessible, and not all particles can be targeted or controller individually. In addition feedback control is very difficult to implement since the characteristic time is usually shorter than that of the available control devices. Although cooperative control has provided analysis methods and synthesis tools that were successfully applied to real-world systems such as autonomous vehicle systems at the macro- and microscales, cooperative control in other application domains such as the nanoscale systems has not been fully explored yet. New real-world problems and application domains pose new challenges for nonlinear cooperative control design.

1.2 Background and Related Work

The book addresses real-world applications of cooperative control in three application domains: networked communication systems, cooperating multi-robotic systems, and multiagent physics systems. In this section, we provide background and related work for each of the application domains.

1.2.1 Networked Communication Systems

Part I studies distributed consensus for networked communication systems. In particular, after presenting average consensus for quantized communication, two emerging applications of distributed consensus in CRNs will be discussed, which include distributed spectrum sensing and radio environment mapping (REM).

CRNs are an innovative approach to wireless engineering in which radios are designed with an unprecedented level of intelligence and agility. This advanced technology enables radio devices to use spectrum (i.e., radio frequencies) in entirely new and sophisticated ways. Cognitive radios have the ability to monitor, sense, and detect the conditions of their operating environment and also dynamically reconfigure their own characteristics to best match those conditions [34].

1.2.1.1 Cooperative Spectrum Sensing

Due to rapidly growing demands of emerging wireless services and new mobile applications for anytime and anywhere connectivity in our daily lives, we expect to face a shortage of wireless spectrum. However, this spectrum-shortage problem is reported to be rooted in the conventional static spectrum-allocation policy where only licensed devices can operate on a designated spectrum band. For example, according to the report from the Shared Spectrum Company, only an average of 5.2% of wireless spectrum under 3 GHz was actively used, indicating that a large fraction of spectrum bands were unutilized or underutilized at any given location and time.

CRNs have emerged as an enabling technology to mitigate the spectrum-scarcity problem. In CRNs, unlicensed (or secondary) devices/users can opportunistically access temporarily available licensed spectrum bands, that is, spectrum bands not being used by the primary users. As a first step toward realization of the new concept of opportunistic spectrum access, the Federal Communications Commission (FCC) has approved the operation of unlicensed cognitive radio (CR) devices in ultra high frequency (UHF) bands (a.k.a. TV white spaces). The first standardization effort based on this CR technology, that is, the IEEE 802.22 wireless regional area networks, is also in its final stage of development [35, 36]. Thus, the openness of the lower-layer protocol stacks in CR and their subsequent ability to adapt their waveforms make them an appealing solution to dynamic spectrum access and alleviate the spectrum-scarcity problem.

Accurate and robust spectrum sensing is essential to spectral efficiency in CRNs. Conventional centralized cooperative spectrum sensing requires that the entire received data be gathered at one place, which may be difficult due to communication constraints [37]. In particular, the multi-hop communication channel requirements of the relay-assisted sensing may bring extra power cost and the sensing data quality may degrade during the multi-hop communication paths. Since future CRNs will consist of heterogeneous devices such as smartphones, tablets, and laptops moving with the collective behaviors of people, consensus-based distributed spectrum sensing [38, 39] reveals great potential for future development of distributed CRNs due to one-hop communication, self-organization, and scalable network structure. Chapter 4 presents a new weighted average consensus approach for distributed spectrum sensing.

1.2.1.2 Radio Environment Mapping

REM is first proposed in Ref. [40]. REM information to the CRNs is like the GPS traffic density map for car drivers. Car drivers with traffic density information can choose better routes to avoid traffic jam. CRNs with REM information will improve the utilization of the dynamic spectrum resources. The REM covers a wide range of functions as an integrated database that provides multidomain environmental information and prior knowledge for CRs, such as the geographical features, available services and networks, locations and activities of neighboring radios, and so on. Among those, one of the most fundamental features is the heat map estimation and tracking, such as power spectral density map estimation [41] or, as an alternative, the channel gain estimation [42] and tracking [43].

There are mainly two different types of methodologies for REM. The traditional method is to detect the existence of signal sources, estimate their number, location and parameters, and then estimate the radio effect they induce in their space (i.e., the field) based on signal propagation models. The second method is referred to as direct methods for field estimation, which is to estimate the field without resorting to source identification. In the book, we adopt the second approach to construct REM efficiently in real time. Most existing work on REM uses centralized methods, where a central data collection and processing machine is available to generate the global radio map. Similar to the centralized spectrum sensing problem, those methods suffer the dependency of reporting channels, bandwidth constraints, and scalability issues [44]. There has been limited work on distributed solutions to the REM problem without a central station. In Chapter 5, we will provide new distributed consensus filter-based methods for distributed cooperative estimation of REM.

1.2.2 Cooperating Autonomous Mobile Robots

Part II considers cooperative control of distributed multirobotic systems. Using multirobotic systems rather than a single robot can have several advantages. For example, collectives of simple robots may be simpler in physical design than a large, complex robot, providing opportunities for systems that are more economical, more scalable, and less susceptible to overall failure. Also, through distributed sensing and action, multirobotic systems have the ability to solve problems that are inherently distributed in space, time, or functionality. Perhaps most important, technologies have advanced to the point where mobile, autonomous robot collectives are technically feasible at reasonable prices. A collection of autonomous robots is described as a swarm [45], a colony [46, 47], a collective [48], or robots exhibiting cooperative behaviors [49]. Most work in cooperative mobile robotics began after the introduction of the new robotics paradigm of behavior-based control [50, 51], which is rooted in biological inspirations. Researchers found it instructive to examine the social characteristics of insects and animals, and to apply the findings to the design of multirobot systems [52]. Interested readers are referred to Refs. [53–55] for reviews on multirobot systems.

Early work in the field of distributed robotics demonstrated the use of the simple local control rules of various biological societies—particularly ants, bees, and birds—to the development of similar behaviors in cooperating robot systems [56, 57]. Such systems can be used in applications including search and rescue [58], satellite clustering [59, 60], formation flight [61], formation flying of spacecraft [62, 63], platoon of underwater vehicle [64], cooperative hunting [65], and mobile sensor network [66]. The control strategies for cooperative robots can be organized into three different approaches: behavior-based approaches, virtual structure approaches, and leader-following approaches.

Behavior-based approaches include the work of Parker [67] and of Balch and Arkin [68]. Parker [67] proposed a software architecture for fault-tolerant multirobot cooperation, which incorporates the use of mathematically modeled motivations within each robot to achieve adaptive action selection. Formation keeping is studied within behavior-based framework using motor schemas by Balch and Arkin [68]. The virtual structure approach was proposed and applied to formations of mobile robots by Lewis and Tan [69]; here, the idea is to force an ensemble of robots to behave as if they were particles embedded in a rigid structure. Virtual structure–based methods are also used by Beard and coauthors in [63] for spacecraft formation control, and by Egerstedt and Hu [70]. Leader-following strategies are reported in Refs. [71–73] and the references therein. In comparing the approaches, coordination is achieved through different types of shared knowledge. In the behavioral-based approach, shared knowledge of the relative configuration states are used to achieve coordination. In the virtual structure approach, coordination is achieved through shared knowledge of the states of the virtual structure. In the leader-following approach, coordination is achieved through shared knowledge of the leader's states.

More recently, a graph theorem–based approach has been suggested and its applications to formation control, rendezvousing, and flocking are studied. Studies using the "consensus" concept includes work in Refs. [23, 74, 75], where stability property of a group of agents is connected to the information flow structure characterized by a communication graph. Similar results are presented independently in Refs. [20, 26], where the dynamics of a unicycle robot is considered. Results in Refs. [76, 77] present general formation control frameworks that apply to robot vehicles with high-order linear dynamics. Various results on cooperative control using graph and system theory–based methods can be found in Refs. [29, 66, 78, 79]. Complementary to graph theory–based methods, Refs. [22a and 27], apply matrix theory–based methods and present cooperative and formation control results for general dynamic vehicle systems in their linear canonical forms.

Despite extensive efforts from the robotics and controls communities to develop distributed control methods solving formation control and other aforementioned multirobot cooperation tasks, new emerging applications call for more advanced robotics control techniques. One of such demands is to respond to the recent Deepwater Horizon oil spill and use ocean robots to detect, monitor, and track the propagation of oil plumes. In Part II, we present new cooperative control methods for multirobot systems to conduct cooperative tasks including source seeking and plume tracking.

1.2.3 Nanoscale Systems and Laser Synchronization

Part III addresses distribute control of multiagent physics systems, including coupled nanoparticle arrays and coupled laser arrays, both of which present unique challenges that have not be addressed before in nonlinear cooperative control.

1.2.3.1 Control of Nanoscale Systems

The integration of physics principles of macroscopic mechanical systems with control mechanisms and control principles has a rich history, enabling new applications and establishing new research directions. Recent technological advances have allowed the scaling of basic mechanical structures to ever smaller dimensions, including the microelectromechanical system (MEMS) technologies that have evolved over the past few decades and the more recent further scaling of mechanical structures to the nanorange. Early work on MEMS devices quickly demonstrated that the physical behaviors of the microscale elements were quite different from those familiar in macroscale components. This led to an intense period of research and experimentation, seeking to establish the physics principles associated with the microscale mechanical elements and to achieve means of controlling the motion of those elements. These studies led to technologies capable of creating microstructures with predicable and controllable behaviors suitable for practical applications.

More recently, technologies have moved into the nanoscale regime, with basic electronic, optoelectronic, sensor, mechanical, and other components enabling many new applications emerging. At these even smaller scales, nanodevices and nanostructures exhibit behaviors different from those seen at the microscale. The behavior of mechanical nanostructures reflects the approaching of structures with a number of atoms, with physics principles migrating into quantum and interacting atom regimes. Ultimately, technologies for these nanostructures will advance to the point where major applications will become routine. However, the application of these nanomechanical structures will require an integration of the new physics principles for their behavior with new control theoretical principles appropriate for the new behaviors exhibited by nanostructures. Similar to the case of MEMS, progress will require an integration of new principles drawing upon emerging experimental results on nanostructures. The availability of new tools and the development of new techniques in handling nanoparticles give unprecedented opportunity for theoretical breakthrough toward controllable nanoscale systems. In particular, scanning probes can control and visualize molecular motion that was not possible before, and nanofabrication technology can now make nanoscale features envisioned for the on-chip infrastructure.

Control at the nanoscale presents many challenges. Due to strict confinement and additional constraints, nanosystems are not readily accessible, and not all particles can be targeted or controlled individually. Also, the system dynamics is highly nonlinear; feedback control may be difficult to implement since the characteristic time may be shorter than that of the available control devices. In addition, particle interaction is intrinsic for nanoscale systems due to the interparticle potentials of various physical origins (e.g., the van del Waals' interactions). Although nonlinear control theory has provided analysis methods and synthesis tools that were successfully applied to dynamic systems at the macro- and microscales, control of nanoscale systems from the perspective of nonlinear control theory has not been adequately addressed yet. Chapter 8 presents control of coupled nanoparticles that slide on a surface, and the control goal is to achieve smooth sliding, thereby reducing friction.

1.2.3.2 Laser Synchronization

For many energy-related applications, it is important to concentrate significant amount of energy in a tiny spot. Lasers are naturally one of the more popular light concentration devices and, consequently, are used as a source for directed energy, space and fiber communication, welding, cutting, fusion, and so on. Lasers are also used for clean energy production and can completely operate on solar energy [80–83].

While for some applications a single high beam quality laser can provide sufficient intensity, for others it is imperative to combine lasers into arrays. A coherent beam combination from N lasers on an array will result in total output intensity that scales as the square number of lasers (N^2). However, a coherent combination requires phase synchronization in lasers and does not naturally occur in laser arrays.

While Kuramoto model describes synchronization behaviors of coupled *phase* oscillators, *coupled laser arrays* have a highly nonlinear model and represent a complex system with both technological [84] and theoretical (Refs. [85–89]) importance. Synchronized or phase-locked state, where all the lasers oscillate at a common frequency *and* with fixed phase relationships, is sought from an applied perspective, because such coherent arrays generate much greater power than a single laser. From a theoretical perspective, laser arrays provide a prime example of a system of coupled limit-cycle oscillators, which connects to explorations of pattern formation and many other topics throughout physics, chemistry, biology, and engineering [90, 91].

Most existing theoretical work on laser arrays assumes identical lasers [92–95]. But real devices are nonidentical and have different intrinsic frequencies [93]. Previous work has studied the system's collective behaviors depending on the parameters of the coupling strength, the pump strength, and the width of the distribution of natural frequencies Refs. [85, 86, 88, 96–98]. For intermediate coupling, the dynamics becomes more complicated. In fact, numerical simulations reveal various unsteady collective states between incoherence and phase locking [96]. Global coupling governed by a complex coupling parameter was investigated for a large system of nonidentical lasers in Ref. [98]. So far, there is no systematic analysis of the complex behaviors exhibited by coupled semiconductor lasers with nonidentical parameters, especially the effects of dynamic coupling and different topologies of the network (neither nearest neighbor nor global) are unknown. In Chapter 9, we present new synchronization results for coupled semiconductor laser arrays using recent distributed cooperative control tools; synchronization conditions are characterized with rigorous mathematical proof.

1.3 Overview of the Book

The rest of the book is organized as follows. Chapter 2 reviews fundamental consensus and consensus filter techniques. The remaining chapters are organized in three parts, each of which represents a specific application domain of distributed cooperative control. Part I considers applications in networked communication systems, and includes three chapters to study average consensus in quantized communication, cooperative spectrum sensing, and distributed REM, respectively. Part II deals with distributed multirobotic systems, and includes two chapters to investigate source seeking and dynamic plume tracking by cooperating mobile robots, respectively. Part III presents distributed cooperative control for multiagent physics systems, and includes two chapters to discuss friction control of nanoparticle arrays and synchronizing coupled laser arrays, respectively. We highlight the contribution of each chapter in the following text.

Chapter 2 introduces the concepts of distributed consensus and consensus filters, which are the cornerstones of modern cooperative control techniques. Brief literature review on both distributed consensus and consensus filters are given. Then, following the introduction of graph theory preliminaries, basic distributed consensus protocols are given in both continuous-time and discrete-time formulations. Distributed consensus filters are also reviewed in both continuous-time and discrete-time formulations for the proportional-integral (PI) average consensus filters, which has better convergence performances than other early consensus filter protocols. It is not our intention to provide a comprehensive review on this rich topic. This chapter presents necessary background for proceeding to the following chapters on applications.

Chapter 3 studies average consensus for directed graphs with quantized communication under fixed and switching topologies. In the presence of quantization errors, conventional consensus algorithms fail to converge and may suffer from an unbounded asymptotic mean square error. Robust consensus algorithms are developed in the chapter to reduce the effect of quantization. Specifically, a robust weighting matrix design is introduced that uses the H_∞ performance index to measure the sensitivity from the quantization error to the consensus deviation. Linear matrix inequalities are used as design tools. The mean-square deviation is proven to converge, and its upper bound is explicitly given in the case of fixed topology with probabilistic quantization. Numerical results demonstrate the effectiveness of this method.

Chapter 4 discusses distributed spectrum sensing in CRNs. Existing distributed consensus-based fusion algorithms only ensure equal gain combining of local measurements, whose performance may be incomparable to various centralized soft combining schemes. Motivated by this fact, practical channel conditions and link failures are considered in the chapter, and new weighted soft measurement combining technique is developed without a centralized fusion center. Following the measurement by its energy detector, each secondary user exchanges its own measurement statistics with its local one-hop neighbors, and chooses the information exchanging rate according to the measurement channel condition, for example, the signal-to-noise ratio. Convergence of the new consensus algorithm is rigorously proved, and it is shown that all secondary users hold the same global decision statistics from the weighted soft measurement combining throughout the network. The chapter also provides distributed optimal weight design under uncorrelated measurement channels. The convergence rate of the consensus iteration is given under the assumption that each communication link has an independent probability to fail, and the upper bound of the iteration number of the ϵ-convergence is explicitly given as a function of system parameters. Simulation results show significant improvement of the sensing performance compared to existing consensus-based approaches, and the performance of the distributed weighted design is comparable to the centralized weighted combining scheme.

Chapter 5 presents distributed estimation and tracking for REM. Compared to existing REM using centralized methods, a distributed solution eliminating the central station is provided for map construction. Based on the random field model of the REM with shadow fading effects, consensus-based Kalman filter design is adopted to estimate and track the temporal dynamic REM variation. The unknown parameters of REM temporal dynamics are estimated by a distributed expectation maximization algorithm that is incorporated with Kalman filtering. The proposed approach features distributed Kalman filtering with unknown system dynamics and achieves dynamic REM recovery without localizing the transmitter. Simulation results show satisfactory performances of the proposed method where spatial correlated shadowing effects are successfully recovered.

Chapter 6 considers the problem of source seeking using a group of mobile robots equipped with sensors for source concentration measurement. In the formulation, the robot team cooperatively estimates the gradient of the source field, moves to the source by tracing the gradient-ascending direction, and keeps a predefined formation in movement. Two control algorithms are presented in the chapter with all-to-all and limited communications, respectively. For the case of all-to-all communication, rigorous analytic analysis proves that the formation center of the robots converges to the source in the presence of estimation errors with a bounded error, the upper bound of which is explicitly given. In the case of limited communication where centralized quantities are not available, distributed consensus filters are used to distributively estimate the centralized quantities, and then embedded in the distributed control laws. Numerical simulations are given to validate the effectiveness of the proposed approaches. Experimental results on the E-puck robot platform demonstrate satisfactory performances in a light source-seeking scenario.

Chapter 7 presents robotic tracking of dynamic plume front modeled by the advection–diffusion equation. Different from existing work purely relying on gradient measurement, the transport model of pollution source is explicitly considered in tracking control design. The problem using a single robot is first studied and solved in an estimation and control framework. It is then extended to the multirobot case in a nearest-neighbor communication topology, where the robots form formation while patrolling along the plume front. The distributed control is scalable to a large number of robots. Simulation results show satisfactory performances of the proposed method.

Chapter 8 studies sliding friction of a nanoparticle array. While the problem is approached by chemical means traditionally, a recent approach has received increasing attention to control the system mechanically to tune frictional responses. In the chapter, feedback control laws is explicitly designed for a one-dimensional particle array sliding on a surface subject to friction. The

Frenkel–Kontorova model describing the dynamics is a nonlinear interconnected system and the accessible control elements are average quantities only.
Local stability of equilibrium points of the uncontrolled system is proved in the
presence of linear and nonlinear particle interactions, respectively. A tracking
control problem is then formulated, whose control objective is for the average
system to reach a designated targeted velocity using accessible elements.
Sufficient stabilization conditions are explicitly derived for the closed-loop
error systems using the Lyapunov theory–based methods. Simulation results
show satisfactory performances. The results can be applied to other physical
systems whose dynamics is described by the Frenkel–Kontorova model.

Chapter 9 considers synchronization of coupled semiconductor lasers modeled by coupled Lang and Kobayashi equations. Decoupled laser stability is first
analyzed, and synchronization conditions of coupled laser dynamics is then
characterized. It is rigorously proven that the coupled system locally synchronizes to a limit cycle under the coupling topology of an undirected connected
graph with equal in-degrees. Graph and systems theory is used in synchronization analysis. The results not only contribute to analytic understanding of
semiconductor lasers but also advance cooperative control by providing a real-
world system of coupled limit-cycle oscillators.

References

1 Breder, C.M. (1954) Equations descriptive of fish schools and other animal
aggregations. *Ecology*, **35**, 361–370.
2 Reynolds, C.W. (1987) Flocks, herds, and schools: A distributed behavioral
model. *Computer Graphics*, **21** (4), 25–34.
3 Kunz, H. and Hemelrijk, C.K. (2003) Artificial fish schools: Collective
effects of school size, body size, and body form. *Artificial Life*, **9**, 237–253.
4 Baldassarre, G., Nolfi, S., and Parisi, D. (2003) Evolving mobile robots able
to display collective behaviors. *Artificial Life*, **9**, 255–267.
5 Brogan, D. and Hodgins, J.K. (1997) Group behaviors for systems with
significant dynamics. *Autonomous Robots*, **4**, 137–153.
6 Mogilner, A. and Edelstein-Keshet, L. (1999) A non-local model for a
swarm. *Journal of Mathematical Biology*, **38**, 534–570.
7 Topaz, C.M. and Bertozzi, A.L. (2004) Swarming patterns in a
two-dimensional kinematic model for biological groups. *SIAM Journal of
Applied Mathematics*, **65** (1), 152–174.
8 Gazi, V. and Passino, K.M. (2003) Stability analysis of swarms. *IEEE Transactions on Automatic Control*, **48** (4), 692–696.
9 Gazi, V. and Passino, K. (2004) Stability analysis of social foraging swarms.
IEEE Transactions on Systems, Man, and Cybernetics–Part B: Cybernetics,
34 (1), 539–557.

10 Wiener, N. (1958) *Nonlinear Problems in Random Theory*, MIT Press, Cambridge, MA.

11 Winfree, A.T. (1967) Biological rhythms and the behavior of populations of coupled oscillators. *Journal of Theoretical Biology*, **16** (1), 15–42.

12 Kuramoto, Y. (1975) Self-entrainment of a population of coupled non-linear oscillators, in *International Symposium on Mathematical Problems in Theoretical Physics*, Lecture Notes in Physics, (ed. H. Arakai), Springer, New York, p. 420.

13 Kuramoto, Y. and Nishikawa, I. (1989) Onset of collective rhythms in large populations of coupled oscillators, in *Cooperative Dynamics in Complex Physical Systems* (ed. H. Takayama), Springer, Berlin.

14 Shimoyama, N., Sugawara, K., Mizuguchi, T., Hayakawa, Y., and Sano, M. (1996) Collective motion in a system of motile elements. *Physical Review Letters*, **76** (20) 3870–3873.

15 Vicsek, T., Czirok, A., Ben-Jacob, E., Cohen, I., and Shochet, O. (1995) Novel type of phase transition in a system of self-driven particles. *Physical Review Letters*, **75** (6), 1226–1229.

16 Strogatz, S.H. (2000) From Kuramoto to Crawford: Exploring the onset of synchronization in populations of coupled oscillators. *Physica D*, **143**, 1–20.

17 Jadbabaie, A., Lin, J., and Morse, A. (2003) Coordination of groups of mobile autonomous agents using nearest neighbor rules. *IEEE Transactions on Automatic Control*, **48** (6), 988–1001.

18 Olfati-Saber, R. and Murray, R.M. (2004) Consensus problems in networks of agents with switching topology and time-delays. *IEEE Transactions on Automatic Control*, **49** (9), 1520–1533.

19 Ren, W., Beard, R.W., and Atkins, E.M. (2005) A survey of consensus problems in multi-agent coordination, in *Proceedings of American Control Conference*, Portland, OR, pp. 1859–1864.

20 Lin, Z., Francis, B., and Maggiore, M. (2004) Local control strategies for groups of mobile autonomous agents. *IEEE Transactions on Automatic Control*, **49** (4), 622–629.

21 Moreau, L. (2005) Stability of multi-agent systems with time-dependent communication links. *IEEE Transactions on Automatic Control*, **50**, 169–182.

22 Qu, Z., Wang, J., and Hull, R.A. (2008) Cooperative control of dynamical systems with application to autonomous vehicles. *IEEE Transactions on Automatic Control*, **53** (4), 894–911.

23 Fax, J.A. and Murray, R.M. (2004) Information flow and cooperative control of vehicle formation. *IEEE Transactions on Automatic Control*, **49** (9), 1465–1476.

24 Khatir, M.E. and Davison, E.J. (2004) Cooperative control of large systems, in *Cooperative Control, Lecture Notes in Control and Information Sciences*,

vol. 309 (eds V. Kumar, N.E. Leonard, and A.S. Moore), Springer-Verlag, Berlin/Heidelberg, pp. 119–136.

25 Ren, W. and Beard, R.W. (2005) Consensus seeking in multiagent systems under dynamically changing interaction topologies. *IEEE Transactions on Automatic Control*, **50** (5), 655–661.

26 Lin, Z., Francis, B., and Maggiore, M. (2005) Necessary and sufficient graphical conditions for formation control of unicycles. *IEEE Transactions on Automatic Control*, **50** (1), 121–127.

27 Qu, Z., Wang, J., and Hull, R. (2005) Leadless cooperative formation control of autonomous mobile robots under limited communication range constraints, in *Cooperative Control and Optimization*, Gainesville, FL.

28 Lafferriere, G., Caughman, J., and Williams, A. (2004) Graph theoretic methods in the stability of veicle formations, in *Proceedings of the American Control Conference*, Boston, pp. 3729–3734.

29 Tanner, H.G., Jadbabaie, A., and Pappas, G.J. (2004) Flocking in teams of nonholonomic agents, in *Cooperative Control, Lecture Notes in Control and Information Sciences*, vol. 309 (eds V. Kumar, N.E. Leonard, and A.S. Moore), Springer-Verlag, Berlin/Heidelberg, pp. 229–239.

30 Qu, Z. (2009) *Cooperative Control of Dynamical Systems: Applications to Autonomous Vehicles*, Springer-Verlag, London.

31 Bai, H. and Arcak, M. (2011) *Cooperative Control Design: A Systematic, Passivity-Based Approach*, Springer, New York.

32 Ren, W. and Cao, Y. (2011) *Distributed Coordination of Multi-agent Networks: Emergent Problems, Models, and Issues*, Springer-Verlag, London.

33 Lewis, F.L., Zhang, H., Hengster-Movric, K., and Das, A. (2014) *Cooperative Control of Multi-agent Systems: Optimal and Adaptive Design Approaches*, Springer, London.

34 Mitola, J. (2000) *An integrated agent architecture for software defined radio*, Ph.D. thesis, The Royal Institute of Technology, KTH.

35 Stevenson, C., Chouinard, G., Lei, Z., Hu, W., Shellhammer, S., and Caldwell, W. (2009) IEEE 802.22: The first cognitive radio wireless regional area network standard. *Communications Magazine, IEEE*, **47** (1), 130–138.

36 IEEE 802.22, Working Group on Wireless Regional Area Networks (WRANs). Http://grouper.ieee.org/groups/802/22/ (accessed September 21, 2016).

37 Mishra, S., Sahai, A., and Brodersen, R. (Istanbul, 2006) Cooperative sensing among cognitive radios, in *IEEE International Conference on Communications*, vol. 4, pp. 1658–1663.

38 Yildiz, M.E., Aysal, T.C., and Barner, K.E. (2009) In-network cooperative spectrum sensing, in *EURASIP European Signal Processing Conference*, Glasgow, Scotland.

39 Li, Z., Yu, F.R., and Huang, M. (2010) A distributed consensus-based cooperative spectrum-sensing scheme in cognitive radio. *IEEE Transaction on Vehicular Technology*, **59** (1), 383–393.

40 Zhao, Y. (2007) *Enabling cognitive radios through radio environment maps*, Ph.D. thesis, Virginia Polytechnic Institute and State University.

41 Bazerque, J. and Giannakis, G. (2010) Distributed spectrum sensing for cognitive radio networks by exploiting sparsity. *IEEE Transaction on Signal Processing*, **58** (3), 1847–1862.

42 Kim, S.J., Dall'Anese, E., and Giannakis, G. (2011) Cooperative spectrum sensing for cognitive radios using Kriged Kalman filtering. *IEEE Journal of Selected Topics in Signal Processing*, **5** (1), 24–36.

43 Dall'Anese, E., Kim, S.J., and Giannakis, G. (2011) Channel gain map tracking via distributed Kriging. *IEEE Transactions on Vehicular Technology*, **60** (3), 1205–1211.

44 Akyildiz, I.F., Lo, B.F., and Balakrishnan, R. (2011) Cooperative spectrum sensing in cognitive radio networks: A survey. *Physical Communication*, **4**, 40–62.

45 Beni, G. and Wang, J. (1989) Swarm intelligence in cellular robotic systems, in *Proceedings of the NATO Advanced Workshop on Robotics and Biological Systems*, Tuscany, Italy.

46 Dorigo, M., Maniezzo, V., and Colorni, A. (1996) The ant system: Optimization by a colony of cooperating agents. *IEEE Transactions on Systems, Man, and Cybernetics*, **26**, 1–13.

47 Agah, A. and Bekey, G. (1997) Phylogenetic and ontogenetic learning in a colony of interacting robots. *Autonomous Robots*, **4**, 85–100.

48 Kube, C.R. and Zhang, H. (1993) Collective robotics: From social insects to robots. *Adaptive Behavior*, **2** (2), 189–219.

49 Parker, L.E. (1993) Designing control laws for cooperative agent teams, in *Proceedings of IEEE International Conference on Robotics and Automation*, Atlanta, GA, pp. 582–587.

50 Arkin, R. (1990) Integrating behavioral, perceptual, and world knowledge in reactive navigation. *Robotics and Autonomous Systems*, **6**, 105–122.

51 Brooks, R.A. (1986) A robust layered control system for a mobile robot. *IEEE Transactions on Robotics and Automation*, **2** (1), 14–23.

52 Reif, J.H. and Wang, H. (1999) Social potential fields: A distributed behavioral control for autonomous robots. *Robotics and Autonomous Systems*, **27**, 171–194.

53 Arai, T., Pagello, E., and Parker, L.E. (2002) Editorial: Advances in multi-robot systems. *IEEE Transactions on Robotics and Automation*, **18** (5), 655–661.

54 Balch, T. and Parker, L.E. (2002) *Robot Teams: From Polymorphism to Diversity*, A K Peters, Natick, MA.

55 Cao, Y.U., Fukunaga, A.S., and Kahng, A.B. (1997) Cooperative mobile robotics: Antecedents and directions. *Autonomous Robotics*, **4** (1), 7–27.

56 Mataric, M.J. (1992) Designing emergent behaviors: From local interactions to collective intelligence, in *Proceedings of the Second International*

Conference on Simulation of Adaptive Behavior (eds J. Meyer, H. Roitblat, and S. Wilson), MIT Press, Honolulu, Hawaii, pp. 432–441.

57 Drogoul, A. and Ferber, J. (1992) From Tom Thumb to the Dockers: Some experiments with foraging robots, in *Proceedings of the Second International Conference on Simulation of Adaptive Behavior* (eds J. Meyer, H. Roitblat, and S. Wilson), MIT Press, Honolulu, Hawaii, pp. 451–459.

58 Murphy, R. (2000) Biomimetic search for urban search and rescue, in *Proceedings of the 2000 IEEE/RSJ International Conference on Intelligent Robots and Systems*, Takamatsu, Japan, pp. 2073–2078.

59 Kang, W. and Yeh, H.H. (2002) Co-ordinated attitude control of multi-satellite systems. *International Journal of Robust and Nonlinear Control*, **12**, 185–205.

60 McInnes, C.R. (1995) Autonomous ring formation for a planar constellation of satellites. *AIAA Journal of Guidance, Control, and Dynamics*, **18** (5), 1215–1217.

61 Giulietti, F., Pollini, L., and Innocenti, M. (2000) Autonomous formation flight. *IEEE Control Systems Magazine*, **20**, 34–44.

62 Mesbahi, M. and Hadaegh, F.Y. (2001) Formation flying control of multiple spacecraft via graphs, matrix inequalities, and switching. *AIAA Journal of Guidance, Control, and Dynamics*, **24**, 369–377.

63 Beard, R.W., Lawton, J., and Hadaegh, F.Y. (2001) A coordination architecture for spacecraft formation control. *IEEE Transactions on Control Systems Technology*, **9** (6), 777–790.

64 Stilwell, D.J. and Bishop, B.E. (2000) Platoons of underwater vehicles. *IEEE Control Systems Magazine*, **20** (6), 45–52.

65 Yamaguchi, H. (2003) A distributed motion coordination strategy for multiple nonholonomic mobile robots in cooperative hunting operations. *Robotics and Autonomous Systems*, **43**, 257–282.

66 Ogren, P., Fiorelli, E., and Leonard, N.E. (2004) Cooperative control of mobile sensor networks: Adaptive gradient climbing in a distributed environment. *IEEE Transactions on Automatic Control*, **40** (8), 1292–1302.

67 Parker, L.E. (1998) ALLIANCE: An architecture for fault tolerant multi-robot cooperation. *IEEE Transactions on Robotics and Automation*, **14**, 220–240.

68 Balch, T. and Arkin, R.C. (1998) Behavior-based formation control for multirobot teams. *IEEE Transactions on Robotics and Automation*, **14** (6), 926–939.

69 Lewis, M.A. and Tan, K.H. (1997) High precision formation control of mobile robots using virtual structures. *Autonomous Robots*, **4**, 387–403.

70 Egerstedt, M. and Hu, X. (2001) Formation constrained multi-agent control. *IEEE Transactions on Robotics and Automation*, **17** (6), 947–951.

71 Wang, P.K.C. (1991) Navigation strategies for multiple autonomous mobile robots moving in formation. *Journal Robotic Systems*, **8** (2), 177–195.

72 Desai, J.P., Ostrowski, J.P., and Kumar, V. (2001) Modeling and control of formations of nonholonomic mobile robots. *IEEE Transactions on Robotics and Automation*, **17**, 905–908.

73 Tanner, H., Pappas, G., and Kumar, V. (2004) Leader-to-formation stability. *IEEE Transactions on Robotics and Automation*, **20** (3), 443–455.

74 Olfati-Saber, R. (2006) Flocking for multi-agent dynamic systems: Algorithms and theory. *IEEE Transactions on Automatic Control*, **51** (3), 401–420.

75 Ren, W., Beard, R.W., and McLain, T.W. (2004) Coordination variable and consensus building in multiple vehicle systems, in *Cooperative Control, Springer-Verlag Series : Lecture Notes in Control and Information Sciences*, vol. 309 (eds V. Kumar, N.E. Leonard, and A.S. Moore), Springer-Verlag, Berlin/Heidelberg, pp. 171–188.

76 Dong, W. and Guo, Y. (2005) Formation control of nonholonomic mobile robots using graph theoretical methods, in *Cooperative Control and Optimization*, Gainesville, FL.

77 Lafferriere, G., Williams, A., Caughman, J., and Veerman, J. (2005) Decentralized control of vehicle formations. *Systems and Control Letters*, **54**, 899–910.

78 Leonard, N.E. and Fiorelli, E. (2001) Virtual leaders, artificial potentials and coordinated control of groups, in *Proceedings of IEEE Conference on Decision and Control*, Orlando, FL, pp. 2968–2973.

79 Lin, J., Morese, A.S., and Anderson, B.D.O. (2003) The multi-agent rendezvous problem, in *Proceedings of IEEE Conference on Decision and Control*, Maui, Hawaiian, pp. 1508–1513.

80 Yabe, T., Bagheri, B., Ohkubo, T., Uchida, S., Yoshida, K., Funatsu, T., Oishi, T., Daito, K., Ishioka, M., Yasunaga, N., Sato, Y., Baasandash, C., Okamoto, Y., and Yanagitani, K. (2008) 100 W-class solar pumped laser for sustainable magnesium-hydrogen energy cycle. *Journal of Applied Physics*, **104** (8), 083104.

81 International Energy Agency, Key World Energy Statistics, 2007, http://www.coprocem.com/documents/key_stats_2007.pdf (accessed September 21, 2016).

82 Ohkubo, T. (2009) Solar-pumped 80 W laser irradiated by a fresnel lens. *Optics Letters*, **34** (2), 175–177.

83 Sakurai, Y., Yabe, T., Ikuta, K., Sato, Y., Uchida, S., and Matsunaga, E. (2008) Basic characterization of the mg combustion engine for a renewable energy cycle using solar-pumped laser. *The Review of Laser Engineering Supplemental Volume*, **36**, 1157–1160.

84 Botez, D. and Scifres, D.R. (1994) *Diode Laser Arrays*, Cambridge University Press, Cambridge, UK.

85 Braiman, Y., Kennedy, T.A.B., Wiesenfeld, K., and Khibnik, A. (1995) Entrainment of solid-state laser arrays. *Physical Review A*, **52** (2), 1500–1506.

86 Fabiny, L., Colet, P., Roy, R., and Lenstra, D. (1993) Coherence and phase dynamics of spatially coupled solid-state lasers. *Physical Review A*, **47** (5), 4287–4296.

87 Garcia-Ojalvo, J., Casademont, J., Torrent, M.C., Mirasso, C.R., and Sancho, J.M. (1999) Coherence and synchronization in diode-laser arrays with delayed global coupling. *International Journal of Bifurcation and Chaos*, **9** (11), 2225–2229.

88 Hohl, A., Gavrielides, A., Erneux, T., and Kovanis, V. (1997) Localized synchronization in two nonidentical coupled lasers. *International Journal of Bifurcation and Chaos*, **23** (25), 4745–4748.

89 Hohl, A., Gavrielides, A., Erneux, T., and Kovanis, V. (1999) Quasiperiodic synchronization for two delay-coupled semiconductor lasers. *Physical Review A*, **59** (5), 3941–3949.

90 Winfree, A.T. (1980) *Geometry of Biological Time*, Springer-Verlag, New York.

91 Kuramoto, Y. (1984) *Chemical Oscillations, Waves and Turbulence*, Springer-Verlag, Berlin.

92 Li, R. and Erneux, T. (1992) Preferential instability in arrays of coupled lasers. *Physical Review A*, **46** (7), 4252–4260.

93 Silber, M., Fabiny, L., and Wiesenfeld, K. (1993) Stability results for in-phase and splay-phase states of solid-state laser arrays. *Journal of the Optical Society of America. B*, **10** (6), 1121–1129.

94 Wang, S.S. and Winful, H.G. (1988) Dynamics of phase-locked semiconductor laser arrays. *Applied Physics Letters*, **52** (21), 1774–1776.

95 Winful, H.G. and Wang, S.S. (1988) Stability of phase locking in semiconductor laser arrays. *Applied Physics Letters*, **53** (20), 1894–1896.

96 Oliva, R.A. and Strogatz, S.H. (2001) Dynamics of a large array of globally coupled lasers with distributed frequencies. *International Journal of Bifurcation and Chaos*, **11** (9), 2359–2374.

97 Zehnle, V. (2000) Theoretical model for coupled solid-state lasers. *Physical Review A*, **63** (03), 033814.

98 Jiang, Z.P. and McCall, M. (1993) Numerical simulation of a large number of coupled lasers. *Journal of the Optical Society of America. B*, **10** (1), 155–163.

2

Distributed Consensus and Consensus Filters

In this chapter, we present the fundamental concepts in multiagent cooperative control, that is, distributed consensus and consensus filters. We first present a very brief literature review on the two concepts and relevant methods and introduce the graph theory preliminaries used in the consensus framework, which is then followed by the mathematical formulation and basic protocols of distributed consensus and consensus filters.

2.1 Introduction and Literature Review

Distributed Consensus

Consensus, also called group agreement, refers to steering a specific variable of all group members to a common value across a multiagent networked system using only local information available to neighboring agents. Consensus is a fundamental concept in multiagent group cooperation behaviors for both natural and humanmade systems, especially for group motion coordination and distributed decision making. For example, the swarm of fishes and fleet of birds maintain certain group shapes or formation, while reaching consensus on velocity and heading angles from time to time, to maximize the group benefits such as reducing energy cost or increasing the group sensing range. Another typical example is platoons of vehicles on highway moving in the same speed while maintaining safe distance between adjacent vehicles. The decision making of a human group or distributed detection of a sensor network depend on reaching consensus on certain decision statistic with global point of view so that the group benefits can be maximized. The common feature of consensus, either in motion coordination or decision making, is that no centralized controller exists for collecting the global information. The global picture for each agent in the network is obtained by information diffusion through local communications among adjacent neighboring agents. The word "distributed" mentioned in this book specifically refers to the local communication mechanism in consensus. Compared to centralized network, distributed network is

Distributed Cooperative Control: Emerging Applications, First Edition. Yi Guo.
© 2017 John Wiley & Sons, Inc. Published 2017 by John Wiley & Sons, Inc.
Companion website: www.wiley.com/go/Guo/DistributedCooperativeControl

more dynamic and scalable, and the function of the network does not rely on certain crucial nodes such as the centralized data fusion center.

The consensus algorithm in electrical systems was first proposed in 1980s for modeling decentralized decision making and parallel computing [1]. The main benefit of consensus is ensuring each node to hold the global average of the initial values throughout the network using local communication between one-hop neighboring nodes. The formulation of consensus builds a base for distributed group coordination [2], and is then widely studied in cooperative control [3] and distributed signal processing [4]. Related work in consensus can be categorized into three main directions as follows:

1. Characterizing network connectivity conditions for convergence analysis of consensus [2, 5–9]
2. Analyzing consensus performances of networked systems with more complicated agent dynamics [10–16]
3. Applying consensus algorithms in various cooperation tasks [17–26]

Early work on consensus in the recent decade is motivated by the biologically inspired distributed group behavior modeling [2], and distributed cooperative robot team design problem [27], for example, robot flag capture game [28]; see the review paper [29]. The graph Laplacian matrix [30] is adopted to model the robotic network [27, 31], which leads to the success of the continuous-time consensus problem formulation [5]. The discrete-time consensus is studied in Refs. [2] and [20] with the aid of nonnegative M-matrix theory [32]. Matrix-based network modeling plays an important role in understanding the consensus algorithm, as it describes the network node-link structure and the information flow of local interactions [2, 5, 7]. One key step in convergence analysis is to utilize spectral properties of graph matrices, that is, Laplacian matrix or M-matrix, to analyze connectivity of the network [5], stability of the consensus iteration [2, 5, 14], the final convergence value [3, 5], the convergence rate [7], and network switching property [8, 9].

Agent has dynamics. Designing consensus-based control laws and analyzing stability of dynamic networked systems form a research line in dynamic cooperative control design [33], which includes several variations of the problem formulation such as formation control [17], synchronization [11], flocking design [34], and swarming [35, 36]. The algorithm has also been extended to double integrator networks [13], higher-order linear systems [14], and passive systems [12]. This research direction extends conventional control design tools to distributed design utilizing local sensing or communications, with the aid of graph matrix analysis in distributed network models. Passivity-based design [37] serves as a typical example, and other interesting problems emerge from biological network synchronization analysis [38, 39].

The research in consensus has shifted from problem formulation, theoretic study, to more practical considerations, such as distributed robust and adaptive design to account for communication delay [40], model uncertainties [41], and dynamic network switching [15]. During the same period, consensus research merges into other fields with development in practical applications, such as environmental sampling [42] and modeling [43]. Besides development in the fields of robotics and controls, consensus algorithms are also studied for distributed detection [21, 26], estimation [20], filtering [19, 22, 23], machine learning [24, 25], and optimization [44, 45], to name a few.

Consensus Filtering

Distributed estimation is a fundamental problem in networked systems. Direct applications of conventional estimation methods often need all-to-all communications, which causes large communication burdens. Much attention has been paid for the past decades to relax the all-to-all communication requirements to neighbor-to-neighbor communications. Although average consensus protocol can be directly applied for the distributed sensor fusion to reach a final estimation with least mean-square errors [20], it does not have explicit input and cannot track the average of time-varying inputs. In more general cases of distributed sensing, each agent has a different input and the goal is to track the average of the set of inputs. In Refs. [19, 46, 47], Olfati-saber and coauthors introduced a distributed low-pass consensus filter and a distributed high-pass consensus filter, which are able to track the average of inputs from all sensors in a network. In the case that the inputs to sensors are not identical, estimation error exists even for a set of constant inputs. Progresses were made to reduce the estimation error, and a proportional-integral (PI) consensus filter is proposed in Refs. [22, 48] to converge accurately to the average of the inputs when the inputs are time-invariant. Examining the PI filter proposed therein, the integral term introduces a zero-zero, which cancels out the zero pole introduced by the constant input. This idea is generalized to ideally track the average of time-varying inputs by exploiting the internal model principle in Ref. [49]. In Ref. [50], the discrete-time counterpart of the PI consensus filter was derived and successfully applied to dynamically merge feature-based maps in robot networks. Modified PI consensus filter is presented in Ref. [51] to remove the requirement of bidirectional exchange of neighboring gains and allow arbitrary switching of network graphs. Further development on the topic includes distributed robust filter with H_∞ consensus measurement [52], information-weighted consensus filter with applications to camera networks [53], and hybrid filters with consensus on information and consensus on measurements [54], to name a few.

Contribution of Our Group

During the past decade, our group has investigated consensus-based cooperative control, paying special attention to the challenges mentioned in Section 1.1.2. Our research focus includes (i) consensus-based cooperative control design addressing system uncertainties and relaxed network conditions, (ii) consensus filtering design for distributed estimation, and (iii) application-oriented cooperative control for various networked multiagent systems. On the front of consensus-based methods, we developed average consensus for quantized communication on directed switching graphs [55], synchronizing uncertain networked Lagrangian systems [56], and adaptive consensus tracking of uncertain multiagent systems [57–59]. Our research results on distributed consensus filters include weighted average consensus filter [60], consensus filtering on directed switching graphs in continuous-time [51] and discrete-time [61] domains. On the application-oriented cooperative control, we developed consensus-based cooperative spectrum sensing of cognitive radio networks [62], consensus filter–based radio environment mapping [63], cooperative source seeking [64] and plume tracking [65] of multirobotic systems, friction control of nanoscale particle arrays [66–69], and semiconductor laser synchronization [70]. While the book presents application-oriented cooperative control, it is worth noting that advances in applications largely depend on theoretical development in the field, especially the consensus and consensus filter–based techniques reviewed earlier.

2.2 Preliminaries on Graph Theory

For a group of networked agents, we assign each agent an index and denote the index set as $\mathcal{I} = \{1, 2, ..., n\}$. We model the n-agent network by a n-node graph and define the graph as $\mathcal{G} := (\mathcal{V}, \mathcal{E})$, where $\mathcal{V} = \{v_i | i \in \mathcal{I}\}$ is the finite nonempty node set, $\mathcal{E} = \{e_{ij} = (v_i, v_j) | i, j \in \mathcal{I}\}$ is the edge set. If the edge set $\mathcal{E}$ is unordered, it is called an *undirected* graph; if $\mathcal{E}$ is ordered, it is called a *directed* graph or a *digraph*. Define $\mathcal{A} = \{a_{ij} | a_{ij} \neq 0 \Leftrightarrow e_{ij} \in \mathcal{E}, a_{ij} = 0 \Leftrightarrow e_{ij} \notin \mathcal{E}\}$ as the adjacent matrix. Particularly, for directed graphs, we refer to v_i and v_j as the tail and head of the edge (v_i, v_j), and assume $a_{ii} = 0$, and $a_{ij} \geq 0$ for all $i \neq j$. The set of neighbors of node v_i is defined as $\mathcal{N}_i = \{v_j | e_{ij} \in \mathcal{E}\}$. We define $\deg_{\text{in}}(v_i) = \sum_{j=1}^{n} a_{ji}$ and $\deg_{\text{out}}(v_i) = \sum_{j=1}^{n} a_{ij}$ as in-degree and out-degree of nodes v_i, respectively. The graph *Laplacian* associated with the digraph $\mathcal{G}$ is defined as

$$\mathcal{L}(\mathcal{G}) = L = \Delta - \mathcal{A}, \tag{2.1}$$

where $\Delta = [\Delta_{ij}]$ is a diagonal matrix with $\Delta_{ij} = 0$ for all $i \neq j$ and $\Delta_{ii} = \sum_{j=1}^{n} a_{ij}$. A digraph $\mathcal{G} = (\mathcal{V}, \mathcal{E})$ is a *balanced graph* if and only if all of its nodes are *balanced*, that is

$$\deg_{\text{in}}(v_i) = \sum_{j} a_{ji} = \sum_{j} a_{ij} = \deg_{\text{out}}(v_i), \quad \forall i \in \mathcal{I}. \tag{2.2}$$

A *directed path* in a digraph is a sequence of edges, $(v_{i_1}, v_{i_2}), \dots, (v_{i_m}, v_{i_{m+1}})$, where $v_{i_j} \in \mathcal{V}$ and $e_{i_j i_{j+1}} \in \mathcal{E}, j = 1, \dots, m$. A directed graph is called *strongly connected* if it is possible to reach any node starting from any other node by traversing edges in the directions in which they point. A *tree* is a graph, where every node, except the root, has exactly one parent node. A *spanning tree* is a tree formed by graph edges that connect all the nodes of the graph. A graph *has a spanning tree* if there exists a spanning tree that is a subset of the graph. For undirected graphs whose edges are unordered, a connected graph is strongly connected and has a spanning tree.

For a finite collection of graphs with a common node set $\mathcal{V}$ and edge set $\mathcal{E}_k, k = 1, \ldots, p$, we call them *jointly connected* if $\left(\mathcal{V}, \bigcup_{k=1}^{p} \mathcal{E}_k\right)$ is a connected graph. Note that for a network of n nodes, the number of all possible communication graphs is finite.

Example 2.1 Figure 2.1 shows a three-node undirected network modeled by an undirected Laplacian matrix, and a three-node directed graphs and its Laplacian matrix. The Laplacian matrix is symmetric for undirected graphs, but not symmetric for directed graphs.

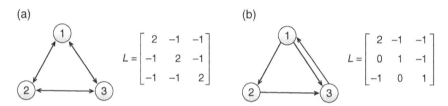

Figure 2.1 Examples of an (a) undirected graph and a (b) directed graph, and their graph Laplacian matrices.

Example 2.2 Figure 2.2 shows a balanced graph, a strongly connected graph, and a graph that has a spanning tree.

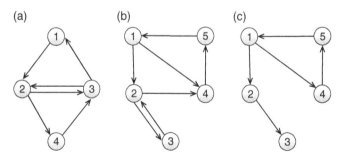

Figure 2.2 Directed graphs with different properties: (a) A balanced graph, (b) a strongly connected graph and (c) a graph that has a spanning tree but is not strongly connected.

Property of Laplacian Matrix

The Laplacian matrix defined in (2.1) is a row-sum-zero matrix and always has a zero eigenvalue with the right eigenvector **1**. We denote the zero eigenvalue as $\lambda_1 = 0$, and its associated eigenvector as $w_r = \mathbf{1} = [1, 1, \ldots, 1]_{1 \times n}^{T}$. We have the following lemma on the Laplacian matrix.

Lemma 2.1 *([30])* Let X be an graph with n vertices and c connected components. Then

$$\text{rank}(L) = \text{rank}(C) = n - c. \tag{2.3}$$

The symmetry of the Laplacian matrix of undirected graphs relates to the graph incidence matrix. If every edge in a digraph has a reverse edge between connected pairs of nodes, then the digraph can be seen as an undirected graph. For undirected graphs, fix an orientation of the graph and let $\mathcal{E}_o = \{e_1, e_2, ..., e_m\}$; we can define the *Incidence matrix* which is an $n \times m$ matrix as $C = [c_{ij}]$ where

$$c_{ik} := \begin{cases} +1 & \text{if } v_i \text{ is the head of the edge } e_k, \\ -1 & \text{if } v_i \text{ is the tail of the edge } e_k, \\ 0, & \text{otherwise.} \end{cases} \tag{2.4}$$

Lemma 2.2 *([30])* The Laplacian matrix of undirected graph $\mathcal{G}$ satisfies $L = CC^T$, regardless of the orientation of $\mathcal{G}$. If the graph is connected, then L is positive semidefinite.

Lemma 2.3 *([5])* A digraph $\mathcal{G} = (\mathcal{V}, \mathcal{E})$ is balanced if and only if every column sum of its Laplacian matrix is zero; in another words, $w_l = \mathbf{1}$ is the left eigenvector of L associated with the zero eigenvalue, that is, $\mathbf{1}^T L = 0$.

Lemma 2.4 *([5])* If a digraph $\mathcal{G} = (\mathcal{V}, \mathcal{E})$ is strongly connected and balanced, having a Laplacian matrix L, denote

$$\hat{L} = \frac{L + L^T}{2} \tag{2.5}$$

then, $\hat{L}$ is a connected undirected graph Laplacian.

Lemma 2.5 *([32])* If a digraph is strongly connected, and

$$w_l = [p_1, ..., p_n] \tag{2.6}$$

is the left eigenvector of the Laplacian matrix L, associated with zero eigenvalue, then all the elements of w_l are nonzero and with the same sign, that is, $p_i > 0, \forall i \in \mathcal{I}$, or $p_i < 0, \forall i \in \mathcal{I}$.

By Lemma 2.5, we choose

$$P = \text{diag}\{p_1, ..., p_n\}, \ p_i > 0, i \in \mathcal{I} \tag{2.7}$$

where p_i is defined in (2.6), and denote

$$\tilde{L} = PL = \begin{bmatrix} p_1 l_{11} & \cdots & p_1 l_{1n} \\ p_2 l_{21} & \cdots & p_2 l_{2n} \\ \vdots & \vdots & \vdots \\ p_n l_{n1} & \cdots & p_n l_{nn} \end{bmatrix} \tag{2.8}$$

It can be easily seen that $\tilde{L}$ is still a graph Laplacian, because it remains row sum zero, diagonal positive, and off-diagonal negative. Moreover, $\tilde{L}$ has a new property that its column sum is also zero. According to Lemma 2.3, $\tilde{L}$ is a Laplacian of a balanced graph with the same set of nodes $\mathcal{V}$ and edges $\mathcal{E}$, but different adjacency weights $\tilde{A}$, $\tilde{a}_{ij} = p_i a_{ij}, \forall a_{ij} \neq 0, i, j \in \mathcal{I}$. Then, from Lemma 2.4,

$$PL + L^T P = \tilde{L} + \tilde{L}^T = 2\hat{L} = 2\hat{C}\hat{C}^T \geq 0, \tag{2.9}$$

where $\hat{L}$ and $\hat{C}$ are the Laplacian and incidence matrix of a connected undirected graph, respectively.

Lemma 2.6 *([71])* If a directed graph has a spanning tree, then its graph Laplacian matrix has a simple zero eigenvalue with an associated eigenvector $\mathbf{1}$, and all of the other eigenvalues are in the open right plane. We denote it as $\mathrm{Re}(\lambda_k) > 0, k = 2, \dots, n$.

We can verify that for a digraph having a spanning tree, the associated Laplacian matrix has rank $n - 1$. Usually, the Laplacian matrices of digraphs are asymmetric. For digraphs with a spanning tree, we can make use of the spectrum properties of the digraph Laplacian matrix to recover the symmetry and to build positive semidefinite matrix.

Lemma 2.7 *([56])* If a digraph $\mathcal{G}$ has a spanning tree and the associated graph Laplacian is L, there exists a symmetric positive definite matrix P satisfying the equation

$$PL + L^T P = Q \geq 0, \tag{2.10}$$

where Q is a positive semidefinite matrix.

Proof: Let J be the Jordan form of L, that is, $L = T^{-1}JT$, where T is a full rank matrix associated with the Jordan transformation. From Ref. [71], if the digraph has a spanning tree, we can choose a suitable T such that $J = \mathrm{diag}\{0, J_1\}$, where $-J_1 \in \mathbb{R}^{(n-1)\times(n-1)}$ is a Hurwitz matrix. According to Theorem 4.6 in Ref. [72], there exists a symmetric positive definite matrix P_1 satisfying $P_1 J_1 + J_1^T P_1 = Q_0 > 0$. Following Theorem 4.29 in Ref. [73], we choose

$$P = T^T \mathrm{diag}\{1, P_1\} T, \tag{2.11}$$

such that

$$PL + L^T P = T^T \begin{bmatrix} 0 & \mathbf{0}_{1\times(n-1)} \\ \mathbf{0}_{(n-1)\times 1} & P_1 J_1 + J_1^T P_1 \end{bmatrix} T$$

$$= T^T \begin{bmatrix} 0 & \mathbf{0}_{1\times(n-1)} \\ \mathbf{0}_{(n-1)\times 1} & Q_0 \end{bmatrix} T \overset{\mathrm{def}}{=} Q \geq 0. \tag{2.12}$$

$\square$

2.3 Distributed Consensus

2.3.1 The Continuous-Time Consensus Protocol

For a network of continuous-time single-integrator agents,

$$\dot{x}_i(t) = u_i(t), \quad i \in \mathcal{I}, \tag{2.13}$$

the consensus protocol is

$$u_i(t) = - \sum_{j \in \mathcal{N}_i} a_{ij}(x_i(t) - x_j(t)), \quad i \in \mathcal{I}, \tag{2.14}$$

where x_i is the state variable of the ith agent, u_i is the control input of the ith agent, a_{ij} is the nonnegative weight constants denoting the coupling strength between the ith agent and the jth agent, $\mathcal{N}_i$ is the set of neighbors of the ith agent and $\mathcal{I} = \{1, \dots, n\}$ is the set of agent index.

In a compact form, we have

$$\dot{x}(t) = u(t) \tag{2.15}$$

$$\text{and} \quad u(t) = -Lx(t), \tag{2.16}$$

where $x = [x_1, \dots, x_n]^T$, $u = [u_1, \dots, u_n]^T$, and L is the Laplacian matrix defined in (2.1).

Consensus under Undirected Graphs

If the communication topology is an undirected graph, L is symmetric and positive semidefinite. We have the Laplacian potential as

$$\Phi(t) = \frac{1}{2}x(t)^T Lx(t). \tag{2.17}$$

The control law (2.16) is naturally the gradient-based feedback control,

$$\dot{x}(t) = -\nabla\Phi(t) = -Lx(t). \tag{2.18}$$

The equilibrium of (2.18) is decided by the spectrum of L. For a connected graph, rank$(L) = n - 1$, and the Laplacian matrix L has a zero eigenvalue corresponding to the right eigenvector $\mathbf{1}$. So $x \in \text{span}(\mathbf{1})$, and we have $\dot{x} \equiv 0$. Thus, the protocol (2.14) achieves consensus, that is,

$$x_i(t) \rightarrow x_j(t), \quad \text{as} \quad t \rightarrow \infty. \tag{2.19}$$

Consensus under Directed Graphs

For general directed graphs, the Laplacian matrix L is not symmetric, (2.17) cannot serve as the potential function. If the directed graph has a spanning tree and the associated graph Laplacian matrix is L, a Lyapunov function can be chosen as

$$V = (Lx)^T P(Lx) = x^T \left(L^T PL\right) x, \tag{2.20}$$

where P is a positive definite matrix defined in (2.11).

Taking derivative of V along the system trajectories of $\dot{x} = -Lx$, we have

$$\dot{V} = x^T \left(-L^T PLL - L^T L^T PL\right) x = -x^T L^T \left(PL + L^T P\right) Lx$$
$$= -(Lx)^T Q(Lx) \leq 0, \tag{2.21}$$

where $Q = PL + L^T P$ is a positive semidefinite matrix defined in (2.12). The system trajectory converges to the set $S = \{x | \dot{V} = 0\} = \{x | Lx = 0\}$. From Lemma 2.6, L has a simple zero eigenvalue with an associated eigenvector $\mathbf{1}$. By LaSalle's invariance principle [72], consensus is achieved, that is, $x_i(t) \to x_j(t)$, as $t \to \infty$.

The consensus result indicates that the agents achieve a group agreement. To decide the final agreement value of the group, we need to look into the graph Laplacian property. It is proved in Ref. [5] that *average* consensus is achieved if and only if $\mathbf{1}^T L = 0$, that is, the graph Laplacian matrix L has a left eigenvector $\mathbf{1}$ associated with the zero eigenvalue. This point can be illustrated in the following text.

Assume $\mathbf{1}^T L = 0$ holds. Since $u(t) = -Lx(t)$ as shown in (2.16), we have

$$\mathbf{1}^T u(t) \equiv 0. \tag{2.22}$$

Since $\mathbf{1}^T \dot{x}(t) = \mathbf{1}^T u(t) \equiv 0$, we get that $\beta = \mathbf{1}^T x$ is an invariant quantity. So,

$$\mathbf{1}^T x^* = \mathbf{1}^T x(0) \tag{2.23}$$

where $x^* = \epsilon \mathbf{1}$, and ϵ is the final agreement value. Thus, we obtain

$$\left(\sum_{i=1}^n 1\right) \epsilon = \mathbf{1}^T x(0) \tag{2.24}$$

$$\Rightarrow \quad \epsilon = \frac{1}{n} \sum_{i=1}^n x_i(0). \tag{2.25}$$

Therefore, average consensus is achieved, that is,

$$x_i(t) \to \frac{1}{n} \sum_{i=1}^n x_i(0), \text{ as } t \to \infty. \tag{2.26}$$

Note that if the graph Laplacian matrix has a nonnegative left eigenvector $\gamma = [\gamma_1, \gamma_2, \ldots, \gamma_n]^T$ associated with the zero eigenvalue, then after reaching consensus, the group agreement value is

$$\epsilon = \frac{\sum_{i=1}^n \gamma_i x_i(0)}{\sum_{i=1}^n \gamma_i}. \tag{2.27}$$

This can be easily seen by replacing $\mathbf{1}$ with γ from (2.22).

Remark 2.1 For a strongly connected graph, P can be simply chosen as $diag(\delta)$, where $\delta^T = [\delta_1, \ldots, \delta_n]$ is the left eigenvector of the Laplacian matrix with the eigenvalue 1, according to Lemma 2.5.

2.3.2 The Discrete-Time Consensus Protocol

For a network of discrete-time single-integrator agents,

$$\dot{x}_i(k+1) = x_i(k) + \alpha u_i(k), \quad i \in \mathcal{I}, \tag{2.28}$$

where $\alpha > 0$ is the stepsize, x_i is the state variable of the ith agent, u_i is the control input of the ith agent, and $\mathcal{I} = \{1, \dots, n\}$ is the set of agent index.

In parallel to the continuous-time protocol of average consensus (2.14), the discrete-time consensus protocol is Ref. ([5])

$$u_i(k) = -\sum_{j \in \mathcal{N}_i} a_{ij}(x_i(k) - x_j(k)), \quad i \in \mathcal{I}. \tag{2.29}$$

where $\mathcal{N}_i(k)$ denotes the set of neighbors of the ith agent at time step k.

Here, we consider a more general *weighted* average consensus protocol [62],

$$x_i(k+1) = x_i(k) + \frac{\alpha}{\delta_i} \sum_{j \in \mathcal{N}_i(k)} (x_j(k) - x_i(k)), i \in \mathcal{I} \tag{2.30}$$

where α is the stepsize satisfying the maximum node degree constraint [3], and $\delta_i \geq 1$ is the weighting ratio to be chosen. If the communication topologies formed by the network are fixed and strongly connected, this weighted average consensus protocol converges to (Ref. [62])

$$x_i(k) \rightarrow x^* = \frac{\sum_{i=1}^{n} \delta_i x_i(0)}{\sum_{i=1}^{n} \delta_i} \text{ as } k \rightarrow \infty, \forall i \in \mathcal{I}. \tag{2.31}$$

Remark 2.2 The choice of the parameter δ_i provides the weights of the final convergence. If δ_i is chosen to be 1 for all i, then (2.30) becomes the same average consensus protocol (2.29).

For convenience, we rewrite the algorithm (2.30) in the following compact form:

$$x(k+1) = Wx(k), \tag{2.32}$$

where $x = [x_1, \dots, x_n]^T$, and W is defined as

$$W = I - \alpha \Delta^{-1} L, \tag{2.33}$$

where $\Delta = \text{diag}\{\delta_1, \dots, \delta_n\}$, $L \in \mathbb{R}^{n \times n}$ is the Laplacian matrix defined in (2.1). The stepsize α satisfies

$$0 < \alpha < \frac{1}{d_{\max}}, \tag{2.34}$$

where $d_{\max} = \max_i l_{ii}$ denotes the maximum node out-degree of the digraph. The convergence of (2.32) depends on the convergence of the infinite matrix product

$$\lim_{k \rightarrow \infty} W^k = \frac{1\delta^T}{\delta^T 1}, \tag{2.35}$$

where $\delta = [\delta_1, \delta_2, \dots, \delta_n]^T$, and δ^T is the left eigenvector of W associated with the eigenvalue 1. We have Theorem 2.1.

Theorem 2.1   For the iteration process (2.30), if the stepsize α satisfies maximum node degree constraint (2.34), and the elements of matrix $\Delta = \mathrm{diag}\{\delta_1, \dots, \delta_n\}$ satisfy $\delta_i \geq 1, \forall i \in I$, and the communication graph is fixed, then the iteration converges to

$$\lim_{k \to \infty} W^k x(0) = \frac{\sum_{i=1}^{n} \delta_i x_i(0)}{\sum_{i=1}^{n} \delta_i} \mathbf{1}. \tag{2.36}$$

That is,

$$x^* = \lim_{k \to \infty} x(k) = \frac{\sum_{i=1}^{n} \delta_i x_i(0)}{\sum_{i=1}^{n} \delta_i}. \tag{2.37}$$

Remark 2.3   Theorem 2.1 is a direct application of the famous Perron–Frobenius theorem [74]. Slightly different versions of this theorem are presented in Refs. [5] and [75]. Setting $\delta_i \geq 1, \forall i \in I$, is a sufficient condition to ensure convergence of the consensus algorithm.

Remark 2.4   Setting $\Delta = I$, all the weight $\delta_i = 1, \forall i$. We have W as a symmetric matrix with real eigenspectrum and eigen space, and 1 is the simple and largest eigenvalue of W. The vector $\mathbf{1}$ and $\mathbf{1}^T$ are the associated left and right eigenvectors, respectively. The convergence of the consensus iteration is given as

$$\lim_{k \to \infty} W^k x(0) = \frac{\mathbf{1}\mathbf{1}^T}{\mathbf{1}^T \mathbf{1}} x(0) = \frac{\sum_{i=1}^{n} x_i(0)}{n} \mathbf{1}, \tag{2.38}$$

which is the average consensus algorithm extensively studied in the literature [7, 20, 21], for example.

2.4   Distributed Consensus Filter

Both the consensus and consensus filters can be used in distributed estimation of sensor networks. Used for distributed estimation, the consensus protocol introduced before can only achieve *static* average consensus, where a snapshot of the input vector is used to initialize the estimator states, after which the input is ignored. In contrast, consensus filters provide *dynamic* average consensus, where the inputs continuously drive the estimators, and the distributed estimator tracks the average of the changing inputs. As reviewed in Section 2.4, different consensus filters have been proposed. We only present the *PI average consensus filters* in this section, which will be used in Chapter 5 to dynamically estimate and track radio environment maps.

2.4.1 PI Average Consensus Filter: Continuous-Time

The PI average consensus filter is first proposed in Ref. [22], which reads as follows:

$$\dot{x}_i(t) = -\gamma x_i(t) - \sum_{j \neq i} a_{ij}(x_i(t) - x_j(t)) + \sum_{j \neq i} b_{ji}(\lambda_i(t) - \lambda_j(t)) + \gamma u_i(t)$$

$$\dot{\lambda}_i(t) = -\sum_{j \neq i} b_{ij}(x_i(t) - x_j(t)). \tag{2.39}$$

Here, $u_i(t) \in \mathbb{R}$ is the input, $x_i(t) \in \mathbb{R}$ is the decision variable, and $\lambda_i(t) \in \mathbb{R}$ is the co-state (i.e., internal estimator state), $\gamma \in \mathbb{R}^+$ is a constant.

The compact matrix form of this protocol writes

$$\dot{x}(t) = -L_p x(t) - \gamma(x(t) - u(t)) + L_I^T \lambda(t)$$

$$\dot{\lambda}(t) = -L_I x(t), \tag{2.40}$$

where $u(t) \in \mathbb{R}^n$, $x(t) \in \mathbb{R}^n$ and $\lambda(t) \in \mathbb{R}^n$ are the input vector, the decision variable vector and the co-state variable vector, respectively. L_p and L_I are Laplacian matrices constructed by $[a_{ij}]$ and $[b_{ij}]$, respectively. Note that both L_I and L_I^T appear on the right side of (2.40) as coefficients (corresponding to the fact that both b_{ij} and b_{ji} appear in (2.43)). This indicates that "weight information must be communicated between agents in addition to the estimator state values" [22].

Define the tracking error vector as follows:

$$e_x(t) = x(t) - \frac{\mathbf{1}\mathbf{1}^T}{n} u(t). \tag{2.41}$$

The convergence property of the protocol (2.39) is proved in Ref. [22] for the following two cases:

1. For any constant input $u \in \mathbb{R}^n$ and any initial states $x(t_0)$, $\lambda(t_0) \in \mathbb{R}^n$, the system states $x(t)$ and $\lambda(t)$ converge to constant vectors and the tracking error $e_x(t)$ approaches to 0 as $t \to \infty$. That is,

$$x_i(t) \to \frac{1}{n} \sum_{i=1}^n u_i, \quad \text{as } t \to \infty, \ \forall i \in I. \tag{2.42}$$

2. For slowly time-varying input $u(t) \in \mathbb{R}^n$, the tracking error $e_x(t)$ is bounded as t increases. The result is consistent to the intuition that we cannot expect to achieve small steady-state tracking errors when the input vector $u(t)$ is changing too rapidly, as it takes time for information diffusion and the effects of $u(t)$ to flow across the network.

2.4.2 PI Average Consensus Filter: Discrete-Time

The continuous-time PI average consensus filter [22] is extended to the discrete-time formulation in Ref. [50] for undirected graph with a fixed topology, which writes

$$x(n + 1) = (1 - h\gamma)x - hLx + hL\lambda(n) + h\gamma u$$

$$\lambda(n + 1) = -hLx(n) + \lambda(n), \tag{2.43}$$

where $u \in \mathbb{R}^m$ is the input vector, $x(n) \in \mathbb{R}^m$ is the decision variable at time n and $\lambda(n) \in \mathbb{R}^m$ is the co-state, L is the Laplacian matrix of the communication graph, $h \in \mathbb{R}^+$ and $\gamma \in \mathbb{R}^+$ are constants satisfying

$$\gamma \geq \frac{3}{2}\lambda_{\max}(L)$$

$$h\gamma < \frac{3}{2}, \tag{2.44}$$

where $\lambda_{\max}(L)$ is the maximum eigenvalue of the Laplacian matrix L.

The consensus filters introduced before provide methods for distributed estimation of global quantities. Compared with the consensus protocols, which does not has an explicit input, protocols (2.39) and (2.43) can track time-varying input signals. Also, compared with some existing consensus filters, such as the low-pass consensus filter [19, 46, 47], protocols (2.39) and (2.43) converge to the average of inputs more accurately. However the following limitations exist in protocols (2.39) and (2.43): (i) These protocols limit to undirected communication graphs with a fixed topology, but does not apply to general directed and switching graphs. (ii) The gain condition in both protocols is subject to knowledge on the maximum eigenvalue of the Laplacian matrix L, which is a global quantity and may be difficult to obtain in practice. These constraints are removed and extensions to general directed switching graphs are made in recent work such as Ref. [51] in a continuous-time formulation and Ref. [61] in a discrete-time formulation.

References

1 Tsitsiklis, J. (1984) *Problems in decentralized decision making and computation*, Ph.D. thesis, Department of Electrical Engineering and Computer Science, M.I.T., Boston, MA.

2 Jadbabaie, A., Lin, J., and Morse, A. (2003) Coordination of groups of mobile autonomous agents using nearest neighbor rules. *IEEE Transactions on Automatic Control*, **48** (6), 988–1001.

3 Olfati-Saber, R., Fax, J., and Murray, R. (2007) Consensus and cooperation in networked multi-agent systems. *Proceedings of the IEEE*, **95** (1), 215–233.

4 Dimakis, A., Kar, S., Moura, J., Rabbat, M., and Scaglione, A. (2010) Gossip algorithms for distributed signal processing. *Proceedings of the IEEE*, **98** (11), 1847–1864.

5 Olfati-Saber, R. and Murray, R. (2004) Consensus problems in networks of agents with switching topology and time-delays. *IEEE Transactions on Automatic Control*, **49** (9), 1520–1533.

6 Moreau, L. (2005) Stability of multi-agent systems with time-dependent communication links. *IEEE Transactions on Automatic Control*, **50** (2), 169–182.

7 Xiao, L. and Boyd, S. (2004) Fast linear iterations for distributed averaging. *Systems & Control Letters*, **53** (1), 65–78.

8 Porfiri, M. and Stilwell, D. (2007) Consensus seeking over random weighted directed graphs. *IEEE Transactions on Automatic Control*, **52** (9), 1767–1773.

9 Tahbaz-Salehi, A. and Jadbabaie, A. (2008) A necessary and sufficient condition for consensus over random networks. *IEEE Transactions on Automatic Control*, **53** (3), 791–795.

10 Chopra, N. and Spong, M. (2005) On synchronization of Kuramoto oscillators, in *Decision and Control, 2005 and 2005 European Control Conference. CDC-ECC '05. 44th IEEE Conference on*, pp. 3916–3922, doi:10.1109/CDC.2005.1582773.

11 Chopra, N. and Spong, M. (2006) Passivity-based control of multi-agent systems. in *Advances in Robot Control: From Everyday Physics to Human-Like Movements*, (eds S. Kawamura and M. Svinin) Springer-Verlag, Berlin/Heidelberg, pp. 107–134.

12 Arcak, M. (2007) Passivity as a design tool for group coordination. *IEEE Transactions on Automatic Control*, **52** (8), 1380–1390.

13 Ren, W. (2008) On concensus algorithms for double-integrator dynamics. *IEEE Transactions on Automatic Control*, **53** (6), 1503–1509.

14 Qu, Z., Wang, J., and Hull, R. (2008) Cooperative control of dynamical systems with application to autonomous vehicles. *IEEE Transactions on Automatic Control*, **53** (4), 894–911.

15 Bai, H. and Arcak, M. (2010) Instability mechanisms in cooperative control. *IEEE Transactions on Automatic Control*, **55** (1), 258–263.

16 Chung, S.J. and Slotine, J.J. (2009) Cooperative robot control and concurrent synchronization of Lagrangian systems. *IEEE Transactions on Robotics*, **25** (3), 686–700.

17 Fax, J. and Murray, R. (2004) Information flow and cooperative control of vehicle formation. *IEEE Transactions on Automatic Control*, **49** (9), 1465–1476.

18 Ogren, P., Fiorelli, E., and Leonard, N. (2004) Cooperative control of mobile sensor networks:adaptive gradient climbing in a distributed environment. *IEEE Transactions on Automatic Control*, **49** (8), 1292–1302.

19 Olfati-Saber, R. and Shamma, J. (2005) Consensus filters for sensor networks and distributed sensor fusion, in *44th IEEE Conference on Decision and Control and 2005 European Control Conference*, Seville, Spain, pp. 6698–6703, doi:10.1109/CDC.2005.1583238.

20 Xiao, L., Boyd, S., and Lall, S. (2005) A scheme for robust distributed sensor fusion based on average consensus. *Proceedings of the Fourth International Information Processing in Sensor Networks*, Los Angeles, CA, pp. 63–70.

21 Kar, S., Aldosari, S., and Moura, J. (2008) Topology for distributed inference on graphs. *IEEE Transactions on Signal Processing*, **56** (6), 2609–2613.

22 Freeman, R., Yang, P., and Lynch, K. (2006) Stability and convergence properties of dynamic average consensus estimators, in *45th IEEE Conference on Decision and Control, 2006*, pp. 338–343, doi:10.1109/CDC.2006.377078.

23 Cortes, J. (2009) Distributed kriged kalman filter for spatial estimation. *IEEE Transactions on Automatic Control*, **54** (12), 2816–2827.

24 Varshavskaya, P., Kaelbling, L., and Rus, D. (2009) Efficient distributed reinforcement learning through agreement, in *Distributed Autonomous Robotic Systems 8* (eds H. Asama, H. Kurokawa, J. Ota, and K. Sekiyama), Springer, Berlin/Heidelberg, pp. 367–378.

25 Forero, P.A., Cano, A., and Giannakis, G.B. (2010) Consensus-based distributed support vector machines. *Journal of Machine Learning Research*, **99**, 1663–1707.

26 Li, Z., Yu, F.R., and Huang, M. (2010) A distributed consensus-based cooperative spectrum-sensing scheme in cognitive radio. *IEEE Transaction on Vehicular Technology*, **59** (1), 383–393.

27 Olfati-Saber, R. and Murray, R.M. (2002) Distributed cooperative control of multiple vehicle formations using structural potential functions, in *Proceedings of the 15th IFAC World Congress*, Barcelona, Spain.

28 D'Andrea, R. and Murray, R. (2003) The roboflag competition, in *American Control Conference, 2003. Proceedings of the 2003*, vol. 1, pp. 650–655, doi:10.1109/ACC.2003.1239093.

29 Murray, R. (2003) Future directions in control, dynamics, and systems: Overview, grand challenges, and new courses. *European Journal of Control*, **9** (2–3), 144–158.

30 Godsil, C. and Royle, G. (2001) *Algebraic Graph Theory, Graduate Texts in Mathematics*, vol. 207, Springer, New York, NY.

31 Fax, J.A. and Murray, R.M. (2002) Graph Laplacians and stabilization of vehicle formations, in *15th IFAC World Congress*, Barcelona, Spain.

32 Bapat, R. and Raghavan, T.E.S (1997) *Nonnegative Matrices and Applications*, Cambridge University Press, Cambridge.

33 Murray, R.M. (2007) Recent research in cooperative control of multivehicle systems. *Journal of Dynamic Systems Measurement and Control*, **129** (5), 571.

34 Olfati-Saber, R. (2006) Flocking for multi-agent dynamic systems: algorithms and theory. *IEEE Transactions on Automatic Control*, **51** (3), 401–420.

35 Sepulchre, R., Paley, D.A., and Leonard, N.E. (2007) Stabilization of planar collective motion: All-to-all communication. *IEEE Transactions on Automatic Control*, **52** (5), 811–824.

36 Sepulchre, R., Paley, D., and Leonard, N. (2008) Stabilization of planar collective motion with limited communication. *IEEE Transactions on Automatic Control*, **53** (3), 706–719.

37 Bai, H., Arcak, M., and Wen, J. (2011) *Cooperative Control Design: A Systematic, Passivity-Based Approach*, Communications and Control Engineering, Springer, New York.

38 Scardovi, L., Arcak, M., and Sontag, E. (2010) Synchronization of interconnected systems with applications to biochemical networks: An input-output approach. *IEEE Transactions on Automatic Control*, **55** (6), 1367–1379.

39 Arcak, M. (2011) Certifying spatially uniform behavior in reaction-diffusion PDE and compartmental ODE systems. *Automatica*, **47** (6), 1219–1229, Special Issue on Systems Biology.

40 Lin, P. and Jia, Y. (2009) Consensus of second-order discrete-time multi-agent systems with nonuniform time-delays and dynamically changing topologies. *Automatica*, **45** (9), 2154–2158.

41 Zhang, W., Wang, Z., and Guo, Y. (2010) Robust consensus for uncertain multi-agent systems on directed communication topologies, in *49th IEEE Conference on Decision and Control (CDC), 2010*, Atlanta, GA, pp. 6317–6322, doi:10.1109/CDC.2010.5717448.

42 Leonard, N., Paley, D., Lekien, F., Sepulchre, R., Fratantoni, D., and Davis, R. (2007) Collective motion, sensor networks, and ocean sampling. *Proceedings of the IEEE*, **95** (1), 48–74.

43 Lynch, K., Schwartz, I., Yang, P., and Freeman, R. (2008) Decentralized environmental modeling by mobile sensor networks. *IEEE Transactions on Robotics*, **24** (3), 710–724.

44 Nedic, A., Ozdaglar, A., and Parrilo, P. (2010) Constrained consensus and optimization in multi-agent networks. *IEEE Transactions on Automatic Control*, **55** (4), 922–938.

45 Boyd, S., Parikh, N., Chu, E., Peleato, B., and Eckstein, J. (2011) Distributed optimization and statistical learning via the alternating direction method of multipliers. *Foundations and Trends in Machine Learning*, **3** (1), 1–122.

46 Olfati-Saber, R. (2006) Distributed Kalman filtering and sensor fusion in sensor networks, in *Networked Embedded Sensing and Control, Lecture Notes in Control and Information Sciences* (eds P. Antsaklis and P. Tabuada), Springer, Berlin/Heidelberg, pp. 157–167.

47 Olfati-Saber, R. (2007) Distributed kalman filtering for sensor networks, in *46th IEEE Conference on Decision and Control*, New Orleans, LA, pp. 5492–5498, doi:10.1109/CDC.2007.4434303.

48 Yang, P., Freeman, R.A., and Lynch, K.M. (2008) Multi-agent coordination by decentralized estimation and control. *IEEE Transactions on Automatic Control*, **53** (11), 2480–2496.

49 Bai, H., Freeman, R.A., and Lynch, K.M. (2010) Robust dynamic average consensus of time-varying inputs, in *49th IEEE Conference on Decision and Control*, Atlanta, GA, pp. 3104–3109.

50 Aragues, R., Cortes, J., and Sagues, C. (2012) Distributed consensus on robot networks for dynamically merging feature-based maps. *IEEE Transactions on Robotics*, **28** (4), 840–854.

51 Li, S. and Guo, Y. (2014) Distributed consensus filter on directed switching graphs. *International Journal of Robust and Nonlinear Control*, doi:10.1002/rnc.3187.

52 Ugrinovskii, V. (2011) Distributed robust filtering with H_∞ consensus of estimates. *Automatica*, **47** (1), 1–13.

53 Kamal, A.T., Farrell, J.A., and Roy-Chowdhury, A.K. (2013) Information weighted consensus filters and their application in distributed camera networks. *IEEE Transactions on Automatic Control*, **58** (12), 3112–3125.

54 Battistelli, G., Chisci, L., Mugnai, G., Farina, A., and Graziano, A. (2015) Consensus-based linear and nonlinear filtering. *IEEE Transactions on Automatic Control*, **60** (5), 1410–1415.

55 Li, S., Guo, Y., Fang, J., and Li, H. (2013) Average consensus with weighting matrix design for quantized communication on directed switching graphs. *International Journal of Adaptive Control and Signal Processing*, **27** (6), 519–540.

56 Zhang, W., Wang, Z., , and Guo, Y. (2014) Backstepping-based synchronization of uncertain networked lagrangian systems. *International Journal of Systems Science*, **45** (2), 145–158.

57 Zhang, W., Wang, Z., and Guo, Y. (2010) Robust consensus for uncertain multi-agent systems on directed communication topologies, in *Proceedings of IEEE International Conference on Decision and Control*, Atlanta, GA, pp. 6317–6322.

58 Wang, Z., Zhang, W., and Guo, Y. (2010) Robust consensus output tracking of multi-agent systems with directed communications, in *Proceedings of ASME Dynamic Systems and Control Conference*, Cambridge, MA, pp. 4193-1–4193-8.

59 Wang, Z., Zhang, W., and Guo., Y. (2011) Adaptive output consensus tracking of a class of uncertain multi-agent systems, in *Proceedings of American Control Conference*, San Francisco, CA, pp. 3387–3392.

60 Li, S. and Guo, Y. (2015) Dynamic consensus estimation of weighted average on a directed graph. *International Journal of Systems Science*, **46** (10), 1839–1853.

61 Li, S. and Guo, Y. (2014) Discrete-time consensus filters on directed switching graphs, in *IEEE International Conference on Control and Automation*, pp. 1096–1101, doi:10.1109/ICCA.2014.6871073.

62 Zhang, W., Wang, Z., Liu, H., Guo, Y., Chen, Y., and Mitola, J. (2015) Distributed consensus-based weight design for cooperative spectrum sensing. *IEEE Transactions on Parallel and Distributed Systems*, **26** (1), 54–64.

63 Kong, R., Zhang, W., and Guo, Y. (2014) Distributed estimation and tracking for radio environment mapping, in *American Control Conference*, pp. 464–470, doi:10.1109/ACC.2014.6859200.

64 Li, S., Kong, R., and Guo, Y. (2014) Cooperative distributed source seeking by multiple robots: Algorithms and experiments. *IEEE/ASME Transactions on Mechatronics*, **19** (6), 1810–1820.

65 Li, S., Guo, Y., and Bingham, B. (2014) Multi-robot cooperative control for monitoring and tracking dynamic plumes, in *IEEE International Conference on Robotics and Automation*, pp. 67–73, doi:10.1109/ICRA.2014.6906591.

66 Guo, Y., Qu, Z., and Zhang, Z. (2006) Lyapunov stability and precise control of the frictional dynamics of a one-dimensional particle array. *Physical Review B*, **73** (9).

67 Guo, Y., Qu, Z., Braiman, Y., Zhang, Z., and Barhen, J. (2008) Nanotribology and nanoscale friction: Smooth sliding through feedback control. *IEEE Control Systems Magazine*, **28** (6), 92–100.

68 Guo, Y. and Qu, Z. (2008) Control of frictional dynamics of a one-dimensional particle array. *Automatica*, **44**, 2560–2569.

69 Guo, Y., Wang, Z., Qu, Z., and Braiman, Y. (2011) Atomic-scale friction control by vibration using friction force microscope. *IFAC Control Engineering Practice*, **19** (11), 1387–1397.

70 Li, S., Guo, Y., and Braiman, Y. (2013) Synchronizing coupled semiconductor lasers under general coupling topologies, in *American Control Conference*, Washington, DC, pp. 1231–1236, doi:10.1109/ACC.2013.6580004.

71 Ren, W. and Beard, R. (2008) *Distributed Consensus in Multi-vehicle Cooperative Control Theory and Applications, Series: Communications and Control Engineering*, Springer-Verlag, London.

72 Khalil, H.K. (2002) *Nonlinear Systems*, Prentice Hall, Upper Saddle River, NJ.

73 Qu, Z. (2009) *Cooperative Control of Dynamical Systems, Application to Autonomous Vehicles*, Springer-Verlag London.

74 Meyer, C. (2000) *Matrix Analysis and Applied Linear Algebra*, SIAM, Philadelphia, PA.

75 Sardellitti, S., Giona, M., and Barbarossa, S. (2010) Fast distributed average consensus algorithms based on advection-diffusion processes. *IEEE Transactions on Signal Processing*, **58** (2), 826.

Part I

Distributed Consensus for Networked Communication Systems

In Part I, we present applications of distributed consensus and consensus filters to networked communication systems. As reviewed in Chapter 2, the average consensus protocol achieves group agreement of a multiagent dynamic system, and the group decision value is the average value of the agent initial states. When applied to networked communication systems, in the presence of quantization (i.e., the process of rounding an analog signal to a digital one), the average consensus protocol cannot guarantee convergence due to the round-off error introduced by quantization. Chapter 3 presents a robust consensus design that achieves average consensus in the presence of quantization error. Utilizing an H_∞ index measurement, the effect of quantization error to the consensus deviation is mitigated by choosing appropriate weighting matrix in the proposed consensus algorithm. Rigorous convergence analysis is given for both fixed and switching directed communication graphs, and the algorithm is easy to implement using existing computational tools for linear matrix inequalities. The proposed quantized average consensus algorithm achieves group agreement for networked communication systems that have sensors and/or actuators with quantization processes.

Following Chapter 3 on quantized consensus, Chapters 4 and 5 present emerging applications of distributed consensus and consensus filters in cognitive radio networks (CRNs). As reviewed in Section 1.2.1, CRNs are intelligent wireless communication systems that optimize the use of available radio frequency (RF) spectrum while minimizing interference to other users. Chapter 4 discusses spectrum sensing, a fundamental problem in CRNs, which aims to improve spectrum utilization by allowing unlicensed secondary users (i.e., smartphone users) to operate in the "white space" of licensed spectrum bands without interfering licensed primary users (i.e., TV towers). Denoting the secondary users as *distributed devices* or *sensors*, our proposed *distributed cooperative sensing* utilizes one-hop communication of distributed devices, and enables them to detect the presence of a primary user in the spectrum. Distributed consensus protocol is used for the purpose of fusing sensor

Distributed Cooperative Control: Emerging Applications, First Edition. Yi Guo.
© 2017 John Wiley & Sons, Inc. Published 2017 by John Wiley & Sons, Inc.
Companion website: www.wiley.com/go/Guo/DistributedCooperativeControl

data over the network and estimating a common sensed variable (which is usually the detected energy level). However, due to imperfect communication channel conditions such as practical multipath fading and shadowing effects, the equal gain combining, which corresponds to the conventional *average* consensus, does not perform well as all sensor measurements are treated equally. In practical CRNs, channel conditions need to be considered, where the sinal-to-noise ratio of the channel can be used to indicate accuracy of each sensor measurement. In this case, a *weighted* average consensus algorithm is needed for improved detection performances of CRNs. The weighted average consensus in discrete-time formulation presented in Chapter 2 is revisited in this chapter, and its application to distributed cooperative spectrum sensing is discussed with adequate details. Simulation results are demonstrated with satisfactory performances.

Chapter 5 presents applications of consensus filters to radio environment mapping (REM), which aims to create coverage mapping of wireless networks based on radio signal strengths received by RF sensors. Assuming RF sensors are uniformly distributed in an area, distributed REM utilizes local communications among neighboring sensors to fuse information over the network for real-time radio map construction. The radio environment is formally modeled as a random field with spatial and temporal dynamics, and the problem is formulated as a distributed estimation problem that can be solved in the framework of Kalman filtering. Due to the lack of centralized information collection, each sensor communicates with its neighbors on important system parameters, and global information is obtained by each sensor through information propagation of the underlying consensus filtering process. As the signal source may be moving, it is a *dynamic* estimation problem; thus, consensus filters are needed rather than static estimation using consensus. The PI average consensus filter in discrete-time formulation reviewed in Chapter 2 is revisited in this chapter, and embedded into the Kalman filter framework to distributively estimate dynamic REM. Simulation results are shown for both static and dynamic radio sources with satisfactory performances.

3

Average Consensus for Quantized Communication

3.1   Introduction

In the past decade, distributed average consensus and gossiping algorithms have received considerable attention in multiple agent coordination and distributed sensor networks including distributed estimation [1–3] and cooperative control [4, 5]. In these studies, it is often assumed that each sensor, say in a wireless sensor network, can communicate with its neighbors without any distortion. This assumption, however, may not be true in practice when considering quantization brought by digital communication. Quantization refers to the process of rounding an analog signal to a digital one to incorporate with digital communications, and the round-off error introduced by quantization is called the "quantization error." As pointed out in Ref. [6], in the presence of quantization errors or noise, the conventional linear consensus algorithm diverge and may have an unbounded asymptotic mean square error. Motivated by this challenge, average consensus with quantized communication has attracted a lot of studies over the past decade.

Based on the convergence of controlled Markov processes, a protocol for undirected topologies is introduced in Ref. [7] to deal with the distributed consensus problem in the presence of quantization errors. This protocol adds a controlled amount of statistical dither before quantization and can reach almost sure (a.s.) convergence to a finite random variable. However, there is no guarantee that this finite random variable is equal to the average of the nodes' initial values. Quantized consensus studied in Refs. [8] and [9] restrict the value of each node to be an integer and reaches average consensus in a "quantized consensus" sense, which means that every node takes one of two neighboring quantization values while preserving the average. However, this is not a strict consensus and nodes reach the quantized average instead of the true average. Average consensus can be asymptotically achieved with the proposed encoding–decoding scheme in Ref. [10], which is based on quantization of scaled innovations. However, this method requires that all nodes are synchronized, which may not be true in a practical network. Ceragioli et al. [11] consider the continuous-time quantized consensus problem, where a

Distributed Cooperative Control: Emerging Applications, First Edition. Yi Guo.
© 2017 John Wiley & Sons, Inc. Published 2017 by John Wiley & Sons, Inc.
Companion website: www.wiley.com/go/Guo/DistributedCooperativeControl

discontinuous protocol is designed to restrict the dynamic evolving toward the initial average. However, there exists unavoidable chattering in practice while replacing the discontinuous function with a hysteresis one indeed remedies the problem but at the cost of accuracy. A protocol featuring an attractive property of preserving the initial average throughout the iterations was proposed in Ref. [12]. This protocol drives the system close to the average consensus with a bounded error. To further reduce the quantization error effects, a sequence averaging algorithm based on this sum-preserving protocol was proposed in Ref. [13]. Meanwhile, [14–16] studied the quantization problem in gossip based consensus algorithms. In Ref. [16], the author proposed two different protocols, that is, a globally quantized strategy and a partially quantized strategy, respectively. For the globally quantized strategy, it was shown that all states eventually reach agreement, but the value of the agreement is different from the initial average. In contrast, the partially quantized strategy preserves the average and the consensus error is bounded by one (states are quantized to integers) after a finite period of time. The partially quantized strategy is also studied in Refs. [14] and [15] and similar conclusions are extended to general quantizers. It was shown that the consensus error of this algorithm is bounded by one, but whether or not the consensus error can be further reduced was not explored.

In this chapter, we consider distributed consensus on directed switching communication graphs, which are used to model many real communication systems [17]. Specifically, inspired by the idea of robust H_∞ consensus control [18], we model the problem as a robust consensus design problem with mismatched quantization noises and design the weighting matrix of the protocol introduced in Ref. [12] under the H_∞ performance index to mitigate the consensus deviation caused by quantization errors. Using linear matrix inequality (LMI) as design tools, we show that the consensus deviation is bounded by a predesignated constant. For a probabilistic quantization scheme on directed graphs, the covariance matrix of states is proven to converge and its steady-state value is given. Moreover, the mean-square deviation from consensus is proved to be upper bounded and the upper bound is explicitly given in terms of the quantization standard deviation and the design parameter. Extensive simulations show the proposed algorithm overperforms existing ones in reducing consensus deviation.

The chapter has the following contributions. First, most studies on average consensus with quantization are limited to undirected graphs [7, 8, 10, 11, 13–16] without allowing arbitrary switching [7–9, 12, 13]. Our results are more general since an undirected graph can be treated as a trivial case of directed graphs. Second, compared to the work [19–21], where weighting matrix design is considered to optimize the convergence rate without a quantizer, we aim at reducing consensus deviation caused by quantization errors, and the upper bound of the mean-square derivation is given for a fixed and directed graph under a probabilistic quantization scheme. Third, compared with the work in Ref. [18], which studies H_∞ consensus control in continuous time with an additive noise satisfying matching conditions, the discrete-time quantized consensus problem is studied in the chapter with a mismatched quantization noise.

Organization: The rest of the chapter is organized as follows. Section 3.2 presents the problem formulation and introduces the protocol of average consensus with quantization. Then, the weighting matrix design method is provided to choose the protocol parameters in Section 3.3. Simulation results are shown in Section 3.4 with performance comparison with existing methods. Finally, Section 3.5 concludes the chapter.

3.2 Problem Formulation

3.2.1 Average Consensus Protocol with Quantization

In the chapter, we study the following consensus protocol with quantization:

$$x_i(t + 1) = x_i(t) + \sum_{j \in N_i} w_{ji}(Q(x_j(t)) - Q(x_i(t))), \tag{3.1}$$

where $W = [w_{ij}]$ is a doubly stochastic matrix, N_i is the neighborhood of node i, and $Q(\cdot)$ is the quantization function (see Appendix C). We can write (3.1) in a matrix equation form as follows:

$$x(t + 1) = Wx(t) - L(Q(x(t)) - x(t)) \tag{3.2}$$

Here, $L = I - W$. This protocol was first proposed in Ref. [12] and independently studied in Ref. [13] because of its attractive property in the fields, such as signal processing and data fusion, where the average of state values is invariant with time. This property can be easily verified by multiplying $\mathbf{1}^T$ on both sides of Equation (3.2).

The presence of quantization function $Q(\cdot)$ in this protocol introduces nonlinearity into the linear consensus model, which makes the new nonlinear model hard to analyze directly as compared to the following conventional consensus protocol:

$$x(t + 1) = Wx(t). \tag{3.3}$$

The difference lies in that the protocol (3.2) introduces an extra nonlinear item $Q(x(t)) - x(t)$. One way to study the new system is to treat this extra item as a disturbance to the nominal linear system (3.3), as shown in the following. This allows us to utilize some tools for linear consensus protocol. Henceforth, we write (3.2) as follows:

$$x(t + 1) = Wx(t) - Lv(t), \tag{3.4}$$

where $v(t) = Q(x(t)) - x(t)$ is the disturbance.

3.2.2 Problem Statement

It is the quantization error in protocol (3.1) that leads to the deviation from the average consensus. This motivates us to utilize a robust control strategy to reduce the effect of the quantization error.

Representing the deviation $x(t) - \frac{11^T x(0)}{n}$ by $z(t)$, we have the following:

$$x(t+1) = Wx(t) - Lv(t)$$

$$z(t+1) = x(t+1) - \frac{11^T x(0)}{n} \tag{3.5}$$

Consider the following H_∞ norm objective function:

$$\|G_{vz}\|_\infty < \gamma, \tag{3.6}$$

where G_{vz} is the transfer function from v to z of (3.5), $\|G_{vz}\|_\infty$ is the H_∞ norm of G_{vz} and γ is a positive constant.

The H_∞ norm performance index (3.6) is defined in the complex frequency domain. It corresponds to the L_2 gain in the time domain; see Appendix D. This inequality imposes a constraint on the L_2 gain from the disturbance $v(t)$ to $z(t)$, which measures the deviation from the average consensus. That is to say, (3.6) is equivalent to the following time domain inequality:

$$\sup_{t>0} \sum_{\tau=0}^{t} \left(z^T(\tau)z(\tau) - \gamma^2 v^T(\tau)v(\tau) \right) < \text{constant}, \tag{3.7}$$

where the constant on the right side of the inequality is due to the effect of the initial states to the output z and is determined by the initial states.

Problem Statement: Design the weighting matrix W of system (3.5) to achieve a bounded consensus error of $x(t)$ and to satisfy the performance index (3.7) (equivalent to the performance index (3.6)).

Remark 3.1   Note that the matrix W is a stochastic matrix. The traditional robust H_∞ control theory is invalid since W has an eigenvalue of 1. This point is analogous to the continuous-time case with a singular Laplacian matrix as the system matrix, to which traditional H_∞ control fails [5, 18].

3.3 Weighting Matrix Design for Average Consensus with Quantization

In this section, we present our main results on the weighting matrix design. We use a linear transformation to transform the original consensus problem to an equivalent stabilization problem. This transformation simplifies the analysis. Based on this transformation, we use a set of LMIs to design the weighting matrices for fixed graphs and switching graphs, respectively. In addition, we study $E(z(t)z^T(t))$, where $E(\cdot)$ denotes the expected value, for the protocol on a fixed graph with the probabilistic quantization scheme, which reveals the statistical property of this method.

3.3.1 State Transformation

We have Proposition 3.1 about the state transformation.

Proposition 3.1 For a strongly connected, balanced, and directed graph, system (3.5) is equivalent to the following decoupled system:

$$y_0(t + 1) = T_0 W T_0^T y_0(t) - T_0 L \upsilon(t)$$
$$y_a(t + 1) = y_a(t)$$
$$z(t + 1) = T_0^T y_0(t + 1) \tag{3.8}$$

under the similarity transformation,

$$\begin{bmatrix} y_0(t) \\ y_a(t) \end{bmatrix} = y(t) = Tx(t), \tag{3.9}$$

where

$$T = \begin{bmatrix} T_0 \\ \frac{1^T}{\sqrt{n}} \end{bmatrix} \tag{3.10}$$

and satisfies $T^T T = T T^T = I$, T_0 is a $(n-1) \times n$ matrix.

Proof: Because

$$I = T T^T = \begin{bmatrix} T_0 T_0^T & T_0 \frac{1}{\sqrt{n}} \\ \frac{1^T}{\sqrt{n}} T_0^T & 1 \end{bmatrix} = T^T T = T_0^T T_0 + \frac{11^T}{n} \tag{3.11}$$

we have,

$$T_0 1 = 0, \, T_0 T_0^T = I, \, T_0^T T_0 = I - \frac{11^T}{n}. \tag{3.12}$$

Representing the system (3.4) by the new variable $y(t)$, we have the following:

$$y(t + 1) = TWT^T y(t) - TL\upsilon(t)$$
$$= \begin{bmatrix} T_0 W T_0^T & T_0 W \frac{1}{\sqrt{n}} \\ \frac{1^T}{\sqrt{n}} W T_0^T & \frac{1^T}{\sqrt{n}} W \frac{1}{\sqrt{n}} \end{bmatrix} y(t) - \begin{bmatrix} T_0 L\upsilon(t) \\ \frac{1^T}{\sqrt{n}} L\upsilon(t) \end{bmatrix}. \tag{3.13}$$

For the strongly connected and balanced graph, W is a doubly stochastic matrix. Therefore, $W1 = 1$ and $1^T W = 1^T$. Since $L = I - W$, we have $1^T L = 1^T - 1^T W = 0$. Together with (3.12), we get

$$y(t + 1) = \begin{bmatrix} T_0 W T_0^T & 0 \\ 0 & 1 \end{bmatrix} y(t) - \begin{bmatrix} T_0 L\upsilon(t) \\ 0 \end{bmatrix}. \tag{3.14}$$

That is,

$$y_0(t+1) = T_0 W T_0^T y_0(t) - T_0 L v(t)$$
$$y_a(t+1) = y_a(t) \tag{3.15}$$

As to the output $z(t)$, we can represent it in terms of $y_0(t)$ as follows:

$$z(t) = x(t) - \frac{\mathbf{11}^T x(0)}{n} = x(t) - \frac{\mathbf{11}^T x(t)}{n} = T_0^T T_0 x(t) = T_0^T y_0(t). \tag{3.16}$$

The transformation matrix T is orthogonal, so the new system with y as the state is equivalent to the original one. This completes the proof. $\qquad \square$

Remark 3.2 The output $z(t)$ only depends on $y_0(t)$, which is not coupled with $y_a(t)$. Therefore, we only need to study the dynamics of $y_0(t)$ in order to study the output $z(t)$.

Remark 3.3 Matrix T is not unique. One example is the matrix with all rows equal to the normalized eigenvectors of matrix $\mathbf{11}^T$ with the eigenvector $\mathbf{1}^T/\sqrt{n}$ in the last row.

Remark 3.4 The dynamic of $y_0(t)$, which completely determines $z(t)$, is asymptotically stable because all the eigenvalues of the system matrix $T_0 W T_0^T$ locates inside the unit circle. This can be observed from the following analysis:

$$TWT^T = \begin{bmatrix} T_0 \\ \frac{\mathbf{1}^T}{\sqrt{n}} \end{bmatrix} W \begin{bmatrix} T_0^T & \frac{1}{\sqrt{n}} \end{bmatrix} = \begin{bmatrix} T_0 W T_0^T & \mathbf{0} \\ \mathbf{0} & 1 \end{bmatrix}.$$

Note that T is an orthogonal matrix, so W and $\begin{bmatrix} T_0 W T_0^T & \mathbf{0} \\ \mathbf{0} & 1 \end{bmatrix}$ are similar matrices, and they have the same eigenvalues. For a strongly connected graph, the associated stochastic matrix has one eigenvalue equal to 1 and the magnitude of all the other eigenvalues is less than 1. Therefore, we conclude that the eigenvalues of $T_0 W T_0^T$ have magnitude less than 1. As a result, the system with $v(t)$ as the input and $z(t)$ as the output is bounded-input bounded-output (BIBO). Therefore, we conclude the output $z(t)$ is bounded due to the boundedness of the disturbance $v(t)$.

3.3.2 Design for Fixed and Directed Graphs

Lemma 3.1 gives a sufficient condition to satisfy the design objective (3.7).

Lemma 3.1 For a strongly connected, balanced, and directed graph with a fixed topology, the system (3.5) meets the objective function (3.7) if there exists a symmetric positive definite matrix $P \in \mathbb{R}^{(n-1) \times (n-1)}$ satisfying

$$\begin{bmatrix} I + T_0 W^T T_0^T P T_0 W T_0^T - P & -T_0 W^T T_0^T P T_0 L \\ -L^T T_0^T P T_0 W T_0^T & -\gamma^2 I + L^T T_0^T P T_0 L \end{bmatrix} < 0. \tag{3.17}$$

Proof: Define the Lyapunov function as follows:

$$V(t) = y_0^T(t)Py_0(t), \tag{3.18}$$

where P is a symmetric positive definite matrix. Denote

$$
\begin{aligned}
J(t) &= \sum_{\tau=0}^{t} z^T(\tau)z(\tau) - \gamma^2 v^T(\tau)v(\tau) \\
&= \sum_{\tau=0}^{t} y_0^T(\tau)T_0 T_0^T y_0(\tau) - \gamma^2 v^T(\tau)v(\tau) \\
&= \sum_{\tau=0}^{t} y_0^T(\tau)y_0(\tau) - \gamma^2 v^T(\tau)v(\tau).
\end{aligned} \tag{3.19}
$$

Note that

$$
\begin{aligned}
V(t+1) - V(0) &= \sum_{\tau=0}^{t} V(\tau+1) - V(\tau) \\
&= \sum_{\tau=0}^{t} y_0^T(\tau+1)Py_0(\tau+1) - y_0^T(\tau)Py_0(\tau).
\end{aligned} \tag{3.20}
$$

Substituting the expression of $y_0(\tau+1)$ into Equation (3.20) and summing it up with (3.19), we get

$$
\begin{aligned}
J(t) + V(t) - V(0) \\
= \sum_{\tau=0}^{t} \Big(&y_0^T(\tau)(T_0 W^T T_0^T PT_0 WT_0^T - P + I)y_0(\tau) \\
&+ v^T(\tau)(-\gamma^2 I + L^T T_0^T PT_0 L)v(\tau) - y_0^T(\tau)T_0 W^T \\
&T_0^T PT_0 Lv(\tau) - v^T(\tau)L^T T_0^T PT_0 WT_0^T y_0(\tau) \Big).
\end{aligned} \tag{3.21}
$$

This can be written in a quadratic form as follows

$$
\begin{aligned}
J(t) + V(t) - V(0) \\
= \sum_{\tau=0}^{t} \big[y_0^T(\tau), v^T(\tau) \big] \\
\cdot \begin{bmatrix} I + T_0 W^T T_0^T PT_0 WT_0^T - P & -T_0 W^T T_0^T PT_0 L \\ -L^T T_0^T PT_0 WT_0^T & -\gamma^2 I + L^T T_0^T PT_0 L \end{bmatrix} \cdot \begin{bmatrix} y_0(\tau) \\ v(\tau) \end{bmatrix}.
\end{aligned}
$$

Since

$$
\begin{bmatrix} I + T_0 W^T T_0^T PT_0 WT_0^T - P & -T_0 W^T T_0^T PT_0 L \\ -L^T T_0^T PT_0 WT_0^T & -\gamma^2 I + L^T T_0^T PT_0 L \end{bmatrix} < 0,
$$

we have

$$J(t) < -V(t) + V(0) \Rightarrow J(t) < V(0)$$

$$\Rightarrow \sup_{t>0} \sum_{\tau=0}^{t} z^T(\tau)z(\tau) - \gamma^2 v^T(\tau)v(\tau) < V(0).$$

Choosing the constant in expression (3.7) not less than $V(0)$ leads to the conclusion. This concludes the proof of Lemma 3.1. □

Lemma 3.1 holds for nonsymmetric W, which corresponds to a directed graph. It gives a sufficient condition to meet the performance index (3.7). The condition is represented by a matrix inequality but not a linear one (recall that $L = I - W$). Therefore, Equation (3.17) is hard to solve directly. We next relax it to an LMI, for which powerful mathematic tools exist and can be used to find solutions [22]. This result is stated in Theorem 3.1.

Theorem 3.1　For a strongly connected, balanced, and directed graph with a fixed topology, system (3.5) has a bounded consensus error and meets the performance index (3.7) if there exists a symmetric positive definite matrix $Q \in \mathbb{R}^{(n-1)\times(n-1)}$ satisfying the following LMI:

$$\begin{bmatrix} -Q & S & -T_0 L & 0 \\ S^T & -Q & 0 & QT_0 \\ -L^T T_0^T & 0 & -\gamma^2 I & 0 \\ 0 & T_0^T Q & 0 & -I \end{bmatrix} < 0, \tag{3.22}$$

where $S = T_0 W T_0^T Q$.

Proof: The proof that system (3.5) has a bounded consensus error follows the argument in Remark 3.4. Now we prove the rest part of Theorem 3.1. Since $\text{diag}([I, Q^{-1}, I, I])$ has full rank, we have

$$\begin{bmatrix} -Q & S & -T_0 L & 0 \\ S^T & -Q & 0 & QT_0 \\ -L^T T_0^T & 0 & -\gamma^2 I & 0 \\ 0 & T_0^T Q & 0 & -I \end{bmatrix} < 0$$

$$\Leftrightarrow \text{diag}([I\, Q^{-1}\, I\, I]) \begin{bmatrix} -Q & S & -T_0 L & 0 \\ S^T & -Q & 0 & QT_0 \\ -L^T T_0^T & 0 & -\gamma^2 I & 0 \\ 0 & T_0^T Q & 0 & -I \end{bmatrix}$$

$$\cdot \text{diag}([I\, Q^{-1}\, I\, I]) < 0$$

$$\Leftrightarrow \left[\begin{array}{ccc|c} -Q & SQ^{-1} & -T_0 L & 0 \\ Q^{-1}S^T & -Q^{-1} & 0 & T_0 \\ -L^T T_0^T & 0 & -\gamma^2 I & 0 \\ \hline 0 & T_0^T & 0 & -I \end{array}\right] < 0. \tag{3.23}$$

Using the Schur complement, we get the following equivalent inequality:

$$\begin{bmatrix} -Q & SQ^{-1} & -T_0L \\ Q^{-1}S^T & -Q^{-1} & 0 \\ -L^TT_0^T & 0 & -\gamma^2I \end{bmatrix} - \begin{bmatrix} 0 \\ T_0 \\ 0 \end{bmatrix}(-I)^{-1}\begin{bmatrix} 0 & T_0^T & 0 \end{bmatrix} < 0$$

$$\Leftrightarrow \left[\begin{array}{c|cc} -Q & SQ^{-1} & -T_0L \\ \hline Q^{-1}S^T & -Q^{-1} + T_0T_0^T & 0 \\ -L^TT_0^T & 0 & -\gamma^2I \end{array} \right] < 0. \tag{3.24}$$

Using the Schur complement to this inequality, we equivalently get the following:

$$\begin{bmatrix} -Q^{-1} + T_0T_0^T & 0 \\ 0 & -\gamma^2I \end{bmatrix}$$
$$- \begin{bmatrix} Q^{-1}S^T \\ -L^TT_0^T \end{bmatrix}(-Q)^{-1}\begin{bmatrix} SQ^{-1} & -T_0L \end{bmatrix} < 0. \tag{3.25}$$

Recalling $T_0T_0^T = I$, $S = T_0WT_0^TQ$, defining $P = Q^{-1}$ and then substituting them into the expression, we can finally get the expression of (3.22), which concludes the proof. □

Remark 3.5 Solving the LMI in (3.22) needs the global information of the graph Laplacian matrix. This step is carried out offline, as done in most of work using optimization methods [6, 18, 20]. Once the weighting matrix is obtained, the consensus protocol (3.1) can be used online to achieve consensus in the presence of quantization errors.

For the probabilistic quantization scheme, we are more interested in the statistical performance of the protocol. We next study the steady-state value of $E(z(t)z^T(t))$, which is the covariance matrix of $x(t)$.

As argued in Ref. [13], the quantization error introduced by the probabilistic quantization scheme can be modeled as a random variable with zero mean and finite variance. The quantization errors are independent across sensors and across different iterations. Let σ^2 denote the quantization error variance of each node, we can express the covariance matrix of $v(t)$ as σ^2I. The deviation from consensus $z(t)$, under the probabilistic quantization scheme, is also a random variable. We have Theorem 3.2 about $E(z(t)z^T(t))$.

Theorem 3.2 For system (3.5) on a strongly connected, balanced, and directed graph with a fixed topology under the probabilistic quantization, the covariance matrix $E(z(t)z^T(t))$ converges to a constant matrix Σ_z, whose vectorized form is given by

$$\text{vec}(\Sigma_z) = (T_0^T \otimes T_0^T)(I - (T_0WT_0^T) \otimes (T_0WT_0^T))^{-1} \cdot ((T_0L)$$
$$\otimes(T_0L))\text{vec}(\Sigma_v), \tag{3.26}$$

where Σ_v is the covariance matrix of $v(t)$.

Proof: According to Equation (3.8), we have

$$y_0(t+1)y_0^T(t+1)$$
$$= T_0 W T_0^T y_0(t) y_0^T(t) T_0 W^T T_0^T + T_0 L v(t) v^T(t) L^T T_0^T$$
$$- T_0 W T_0^T y_0(t) v^T(t) L^T T_0^T - T_0 L v(t) y_0^T(t) T_0 W^T T_0^T. \quad (3.27)$$

Calculating the expected value on both sides, we get

$$E(y_0(t+1)y_0^T(t+1))$$
$$= T_0 W T_0^T E(y_0(t) y_0^T(t)) T_0 W^T T_0^T + T_0 LE(v(t) v^T(t)) L^T T_0^T. \quad (3.28)$$

Defining a new variable $\Sigma_{y0}(t) = E(y_0(t)y_0^T(t))$ and noting that $E(v(t)v^T(t)) = \Sigma_v$, which is a constant matrix, we have the following:

$$\Sigma_{y0}(t+1) = T_0 W T_0^T \Sigma_{y0}(t) T_0 W^T T_0^T + T_0 L \Sigma_v L^T T_0^T. \quad (3.29)$$

Use the Kronecker product to write $\Sigma_{y0}(t+1)$ in a vector form as follows:

$$\text{vec}(\Sigma_{y0}(t+1))$$
$$= ((T_0 W T_0^T) \otimes (T_0 W T_0^T))\text{vec}(\Sigma_{y0}(t)) + ((T_0 L) \otimes (T_0 L))\text{vec}(\Sigma_v).$$

This is a linear time-invariant system; therefore, we can study its system matrix for the convergence. The eigenvalue of $(T_0 W T_0^T) \otimes (T_0 W T_0^T)$ is $\lambda_i \lambda_j$ for $i = 1, 2, ..., n, j = 1, 2, ..., n$, where λ_i is the ith eigenvalue of $T_0 W T_0^T$. In addition, $|\lambda_i| < 1$ as argued in Remark 3.4. Therefore, $|\text{eig}((T_0 W T_0^T) \otimes (T_0 W T_0^T))| < 1$. Hence, $\Sigma_{y0}(t)$ converges eventually. Denoting Σ_{y0} as its steady-state value, we can calculate it easily by solving the steady-state linear equation. After calculation, we have the following:

$$\text{vec}(\Sigma_{y0}) = (I - (T_0 W T_0^T) \otimes (T_0 W T_0^T))^{-1}((T_0 L) \otimes (T_0 L))\text{vec}(\Sigma_v). \quad (3.30)$$

Together with $\Sigma_z = T_0^T \Sigma_{y0} T_0$, we get the following:

$$\text{vec}(\Sigma_z) = (T_0^T \otimes T_0^T)\text{vec}(\Sigma_{y0})$$
$$= (T_0^T \otimes T_0^T)(I - (T_0 W T_0^T) \otimes (T_0 W T_0^T))^{-1}((T_0 L) \otimes (T_0 L))\text{vec}(\Sigma_v).$$

This concludes the proof. □

Theorem 3.3 For system (3.5) on a strongly connected, balanced, and directed graph with a fixed topology under the probabilistic quantization, if the performance index (3.7) is satisfied, the mean-square deviation $E(z^T(t)z(t)/n)$ converges and is eventually upper bounded by $\gamma^2 \sigma^2$. That is,

$$\lim_{t \to \infty} E\left(\frac{z^T(t)z(t)}{n}\right) < \gamma^2 \sigma^2, \quad (3.31)$$

where $E(z^T(t)z(t)/n)$ is the expected value of $z^T(t)z(t)/n$, n is the number of nodes in the graph, σ^2 is the quantization error variance of each node, and γ is defined in the performance index (3.7).

Proof: Since $E(z(t)z^T(t))$ converges (Theorem 3.2), we conclude

$$\lim_{t \to \infty} E\left(\frac{z^T(t)z(t)}{n}\right)$$

exists by noting that $z^T(t)z(t) = \text{trace}(z(t)z^T(t))$. According to the performance index (3.7), we have for all $t > 0$,

$$\sum_{\tau=0}^{t} z^T(\tau)z(\tau) - \gamma^2 v^T(\tau)v(\tau) < \text{constant.} \tag{3.32}$$

Calculating expected value on both sides and rewriting the expression, we get

$$\sum_{\tau=0}^{t} E(z^T(\tau)z(\tau)) < t\gamma^2 \, \text{trace}(E(v(\tau)v^T(\tau))) + \text{constant.} \tag{3.33}$$

For the probabilistic quantization, the covariance matrix of v is $\sigma^2 I$. Thus, we have $\text{trace}(E(v(\tau)v^T(\tau))) = n\sigma^2$. Substituting this expression into the inequality (3.33), dividing both side of the inequality with t and then calculating the limit as t goes to infinity, we have

$$\lim_{t \to \infty} E\left(\frac{z^T(t)z(t)}{n}\right) < \gamma^2 \sigma^2 + \lim_{t \to \infty} \frac{\text{constant}}{t}. \tag{3.34}$$

Since $\lim_{t \to \infty} \frac{\text{constant}}{t} = 0$, we get the inequality (3.31). This concludes the proof.

□

Remark 3.6   We can consider $\mathbf{z}(t)$ as the error from the average consensus. Then, $\sqrt{E((\mathbf{z}^T(t)\mathbf{z}(t))/n)}$ is the standard deviation of this consensus error. This theorem reveals that the error sensitivity is less than γ in the sense of standard deviation.

Remark 3.7   Equation (3.31) can be used as a guidance to determine the value of γ for the case with fixed topologies. In practice, we may want the steady-state mean-square deviation $\lim_{t \to \infty} E((z^T(t)z(t))/n)$ upper bounded (say the desired upper bound is α_{ss}). According to Equation (3.31), we can choose $\gamma = \sqrt{(\alpha_{ss})}/\sigma$ in the design. All together, we have the following weighting matrix design procedures for a given strongly connected topology:

1. *Balance Regulation*: Define a $n \times n$ matrix $W = [w_{ij}]$ with $w_{kl} = 0$ if there is no edge from the kth node to the lth one and solve equality constraints $W\mathbf{1} = \mathbf{1}$ and $W^T\mathbf{1} = \mathbf{1}$ to obtained a doubly stochastic W in terms of $p - 2n$ independent scalar variables (p is the number of directed edges on the graph).

2. *LMI Solving*: Choose $\gamma = \sqrt{(\alpha_{ss})}/\sigma$ in the LMI (3.22) and solve a feasible W.

3.3.2.1 Special Case: Undirected Graph

For a strongly connected undirected graph, the weighting matrix is a symmetric one. Due to the special structure, in this case the steady-state standard deviation of the consensus error $\lim_{t \to \infty} E(\mathbf{z}^T(t)\mathbf{z}(t))$ can be explicitly expressed in terms of the spectrum of the graph.

For system (3.5) on a strongly connected undirected graph, which is a special case of balanced and directed graphs, we have the following results according to Theorem 3.2:

$$\lim_{t \to \infty} E(\mathbf{z}^T(t)\mathbf{z}(t)) = \lim_{t \to \infty} \text{trace}(E(\mathbf{z}(t)\mathbf{z}^T(t)))$$

$$= \text{trace}(\Sigma_z) = \text{vec}^T(I)\text{vec}(\Sigma_z)$$

$$= \text{vec}^T(I)(T_0^T \otimes T_0^T)I$$

$$- (T_0 W T_0^T) \otimes (T_0 W T_0^T))^{-1}((T_0 L) \otimes (T_0 L))\text{vec}(\Sigma_v). \tag{3.35}$$

With the Kronecker product property that $(B^T \otimes A)\text{vec}(X) = \text{vec}(AXB)$ for matrices A, B, and X of appropriate sizes, we have $\text{vec}^T(I)(T_0^T \otimes T_0^T) = \text{vec}^T(T_0 T_0^T) = \text{vec}^T(I)$ and $((T_0 L) \otimes (T_0 L))\text{vec}(\Sigma_v) = \text{vec}(T_0 L \Sigma_v L^T T_0^T)$. Also noting that $\Sigma_v = \sigma^2 I$ for the identically independent quantization with the variance σ^2 for each node, we have

$$\lim_{t \to \infty} E(\mathbf{z}^T(t)\mathbf{z}(t))$$

$$= \sigma^2 \text{vec}^T(I)(I - (T_0 W T_0^T) \otimes (T_0 W T_0^T))^{-1}\text{vec}(T_0 L L^T T_0^T). \tag{3.36}$$

The matrix $(T_0 W T_0^T) \otimes (T_0 W T_0^T)$ is symmetric for undirected graph with $W = W^T$, and its eigenvalues locate strictly in the unit circle as pointed out in the proof of Theorem 3.2. Thus, we can express $(I - (T_0 W T_0^T) \otimes (T_0 W T_0^T))^{-1}$ in the following geometric series:

$$(I - (T_0 W T_0^T) \otimes (T_0 W T_0^T))^{-1} = \sum_{j=0}^{\infty} ((T_0 W T_0^T) \otimes (T_0 W T_0^T))^j$$

$$= \sum_{j=0}^{\infty} (T_0 W T_0^T)^j \otimes (T_0 W T_0^T)^j. \tag{3.37}$$

Therefore,

$$\lim_{t \to \infty} E(z^T(t)z(t))$$

$$= \sigma^2 \text{vec}^T(I) \left(\sum_{j=0}^{\infty} (T_0 W T_0^T)^j \otimes (T_0 W T_0^T)^j \right) \text{vec}(T_0 L L^T T_0^T)$$

$$= \sigma^2 \sum_{j=0}^{\infty} \left(\text{vec}^T(I)((T_0 W T_0^T)^j \otimes (T_0 W T_0^T)^j)\text{vec}(T_0 L L^T T_0^T) \right)$$

$$= \sigma^2 \sum_{j=0}^{\infty} \left(\text{vec}^T(I) \text{vec}((T_0 W T_0^T)^j T_0 L L^T T_0^T (T_0 W T_0^T)^j) \right)$$

$$= \sigma^2 \sum_{j=0}^{\infty} \text{trace} \left(I (T_0 W T_0^T)^j T_0 L L^T T_0^T \cdot (T_0 W T_0^T)^j \right)$$

$$= \sigma^2 \sum_{j=0}^{\infty} \text{trace} \left((T_0 W T_0^T)^j T_0 L L^T T_0^T \cdot (T_0 W T_0^T)^j \right). \tag{3.38}$$

By exploiting the property that $T_0 \mathbf{1} = \mathbf{0}$, $T_0 T_0^T = I$ and $T_0^T T_0 = I - \frac{\mathbf{1}\mathbf{1}^T}{n}$, we get $(T_0 W T_0^T)^j = T_0 (W^j - \frac{\mathbf{1}\mathbf{1}^T}{n}) T_0^T$, and further $(T_0 W T_0^T)^j T_0 L L^T T_0^T (T_0 W T_0^T)^j = T_0 (W^j - W^{j+1})^2 T_0^T$ after some trivial calculations. Note that

$$T(W^j - W^{j+1})^2 T^T = \begin{bmatrix} T_0 (W^j - W^{j+1})^2 T_0^T & \mathbf{0} \\ \mathbf{0} & \mathbf{0} \end{bmatrix}, \tag{3.39}$$

where T is defined in (3.10). As $TT^T = I$, equation (3.39) implies the $(n-1) \times (n-1)$ matrix $T_0 (W^j - W^{j+1})^2 T_0^T$ has the same spectrum as the $n \times n$ matrix $(W^j - W^{j+1})^2$ except the zero eigenvalue, that is, matrix $T_0 (W^j - W^{j+1})^2 T_0^T$ has eigenvalues at $(\lambda_i^j - \lambda_i^{j+1})^2$ for $i = 2, 3, ..., n$ with λ_i denoting the ith eigenvalue of W. Therefore,

$$\text{trace} \left((T_0 W T_0^T)^j T_0 L L^T T_0^T \cdot (T_0 W T_0^T)^j \right) = \sum_{i=2}^{n} (\lambda_i^j - \lambda_i^{j+1})^2$$

$$= \sum_{i=2}^{n} \lambda_i^{2j} (1 - \lambda_i)^2. \tag{3.40}$$

Thus,

$$\lim_{t \to \infty} E(z^T(t) z(t)) = \sigma^2 \sum_{j=0}^{\infty} \sum_{i=2}^{n} \lambda_i^{2j} (1 - \lambda_i)^2$$

$$= \sigma^2 \sum_{i=2}^{n} \left((1 - \lambda_i)^2 \left(\sum_{j=0}^{\infty} \lambda_i^{2j} \right) \right)$$

$$= \sigma^2 \sum_{i=2}^{n} \left((1 - \lambda_i)^2 \frac{1}{1 - \lambda_i^2} \right) = \sigma^2 \sum_{i=2}^{n} \frac{1 - \lambda_i}{1 + \lambda_i}. \tag{3.41}$$

Consequently, we conclude that the inequality (3.31), for strongly connected undirected graph, is guaranteed by the following condition:

$$\frac{1}{n} \sum_{i=2}^{n} \frac{1 - \lambda_i}{1 + \lambda_i} < \gamma^2. \tag{3.42}$$

Here, λ_i denotes the ith eigenvalue of W.

For an undirected graph with identically weighted edges, there is only one degree of freedom for the weighting matrix, and it is of the form $W = I - kL_0$ where L_0 is the Laplacian matrix of the graph with unit edge weights. The weighting matrix design criteria is guaranteed by Theorem 3.4.

Theorem 3.4   For system (3.5) on a strongly connected, identically weighted undirected graph with a fixed topology under the probabilistic quantization, the mean-square deviation $E(z^T(t)z(t)/n)$ is eventually upper bounded by $\gamma^2\sigma^2$ under the following condition:

$$0 < k < \frac{2}{\mu_n}min\left\{\frac{n\gamma^2}{n\gamma^2 + n - 1}, 1\right\} \tag{3.43}$$

for $W = I - kL_0$ where L_0 is the Laplacian matrix of the graph with unit edge weights and μ_n is its largest eigenvalue.

Proof: The condition $k < \frac{2}{\mu_n}$ in (3.43) guarantees non-one eigenvalues of W locate strictly inside the unit circle and system (3.5) will not diverge under this condition. Now, to conclude the result, we show $k < \frac{2n\gamma^2}{\mu_n(n\gamma^2+n-1)}$ in inequality (3.43) is sufficient for (3.42).

$$\sum_{i=2}^{n}\frac{1-\lambda_i}{1+\lambda_i} < n\gamma^2 \Leftarrow \sum_{i=2}^{n}\left(-1+\frac{2}{1+\lambda_i}\right) < n\gamma^2$$

$$\Leftarrow (n-1)\left(-1+\frac{2}{2-k\mu_n}\right) < n\gamma^2$$

$$\Leftarrow k < \frac{2n\gamma^2}{\mu_n(n\gamma^2 + n - 1)}. \tag{3.44}$$

Note that $\lambda_i \geq 1 - k\mu_n$ for $i = 2, 3, ..., n$ is used in the derivation of the second step of (3.44). Therefore, (3.44) completes the proof. □

Remark 3.8   According to the Perron–Frobenius theorem [23], μ_n is upper bounded by $2deg_{max}$ with deg_{max} denoting the maximum degree of L_0. As a result, (3.43) holds for $0 < k < \frac{1}{deg_{max}}min\left\{\frac{n\gamma^2}{n\gamma^2+n-1}, 1\right\}$.

Remark 3.9   The largest eigenvalue μ_n can be analytically obtained for some special topologies [24], such as complete graph, cycle topology, and star topology. In these cases, the upper bound of k in (3.43) can be obtained explicitly, as summarized in Table 3.1.

3.3.3   Design for Switching and Directed Graphs

In some practices, failure of previous communication links and emergence of new communication connections may happen. Both will lead to a change of the network topology. Based on this consideration, we study directed graphs with

Table 3.1 The range of k for special topologies with the performance index γ.

Topology	k
Complete graph	$0 < k < \frac{2}{n}\min\{\frac{n\gamma^2}{n\gamma^2+n-1}, 1\}$
Strongly regular graph srg(n,d,a,b)	$0 < k < \frac{4\min\{\frac{n\gamma^2}{n\gamma^2+n-1}, 1\}}{2n-(a-b)+\sqrt{(a-b)^2+4(d-b)}}$
Cycle	$0 < k < \frac{1}{2}\min\{\frac{n\gamma^2}{n\gamma^2+n-1}, 1\}$
Star	$0 < k < \frac{2}{n}\min\{\frac{n\gamma^2}{n\gamma^2+n-1}, 1\}$
Path	$0 < k < \frac{1}{2}\min\{\frac{n\gamma^2}{n\gamma^2+n-1}, 1\}$
Complete bipartite graph	$0 < k < \frac{2}{n}\min\{\frac{n\gamma^2}{n\gamma^2+n-1}, 1\}$

The weighting matrix is chosen to be $W = I - kL_0$ with L_0 denoting the Laplacian matrix of the corresponding graph with unit edge weights.

switching topologies in this section. In this situation, we model the system as a switched system [25], where the switch is triggered by a switching signal $s(t)$. In order to distinguish the difference of weighting matrix W associated with different topologies, in this section we use $G_{s(t)}$, $W_{s(t)}$, and $L_{s(t)}$ to represent the graph, weighting matrix, and Laplacian matrix at time t. For a directed graph with a switching topology, we have similar results to Lemma 3.1 and Theorem 3.1.

Lemma 3.2   For a strongly connected, balanced, and directed graph with a switching topology $G_{s(t)}$, system (3.5) satisfies the objective function (3.7) if there exists a common symmetric positive definite matrix $P \in \mathbb{R}^{(n-1)\times(n-1)}$, satisfying

$$\begin{bmatrix} \Gamma_{11} & \Gamma_{12} \\ \Gamma_{21} & \Gamma_{22} \end{bmatrix} < 0, \tag{3.45}$$

where

$$\Gamma_{11} = I + T_0 W_{s(t)}^T T_0^T P T_0 W_{s(t)} T_0^T - P \tag{3.46}$$

$$\Gamma_{12} = -T_0 W_{s(t)}^T T_0^T P T_0 L_{s(t)} \tag{3.47}$$

$$\Gamma_{21} = -L_{s(t)}^T T_0^T P T_0 W_{s(t)} T_0^T \tag{3.48}$$

$$\Gamma_{22} = -\gamma^2 I + L_{s(t)}^T T_0^T P T_0 L_{s(t)} \tag{3.49}$$

and $s(t)$ denotes the switching signal.

Proof: The proof of this lemma is similar to Lemma 3.1. The difference is that both $W_{s(t)}$ and $L_{s(t)}$ share a common Lyapunov function $V(t) = y_0^T(t)Py_0(t)$. ☐

Theorem 3.5   For a strongly connected, balanced, and directed graph with a switching topology $G_{s(t)}$, system (3.5) has a bounded consensus error and achieves the performance index (3.7) if there exists a common symmetric positive definite matrix $Q \in \mathbb{R}^{(n-1)\times(n-1)}$, satisfying the following LMI:

$$\begin{bmatrix} -Q & S_{s(t)} & -T_0 L_{s(t)} & 0 \\ S_{s(t)}^T & -Q & 0 & QT_0 \\ -L_{s(t)}^T T_0^T & 0 & -\gamma^2 I & 0 \\ 0 & T_0^T Q & 0 & -I \end{bmatrix} < 0. \tag{3.50}$$

Here, $S_{s(t)} = T_0 W_{s(t)} T_0^T Q$ and $s(t)$ denotes the switching signal.

Proof: This theorem follows from Lemma 3.2 and can be similarly proved as Theorem 3.1.   □

For graphs with fixed topologies, we studied the steady-state covariance matrix $E(z(t)z^T(t))$ of the deviation in Theorem 3.3. However, the steady-state value may not exist for the case on a switching graph due to the topology changes. Therefore, no conclusions on the steady-state value is made for the switching case. Nevertheless, Theorem 3.5 guarantees average consensus with an H_∞ performance for the switching topology.

3.4 Simulations and Performance Evaluation

In this section, we give three simulation examples: the first one on a fixed and directed graph, the second one on a directed switching graph, and the third one on an fixed and undirected graph. As shown in Refs. [7] and [13], the probabilistic quantization, compared to the deterministic quantization, has the advantage to reduce the consensus disagreement in a statistical sense. In order to demonstrate that the proposed method is able to further reduce the consensus disagreement, we use the probabilistic quantization scheme in the simulation. Nevertheless, the performance index for the deterministic quantization is also guaranteed according to Theorems 3.1 and 3.5. We compare our proposed method with the maximum-degree weights [12], the LMSC method [6], and the modified LMSC method in the simulation.

3.4.1 Fixed and Directed Graphs

In the first example, we study the case of a fixed and directed graph. There are 10 nodes in the network and the information flow of this network is shown in Figure 3.1. In the simulation, we set $\eta = 2$ and $b = 2$ for the quantizer. We choose $\gamma = 1.1$. The LMI is constructed based on Theorem 3.1 and solved using MATLAB. The initial values of each node are generated according to the Gaussian distribution with zero mean and unit variance. We measure the performance using $z^T(t)z(t)/n$. Results are averaged over 500 Monte Carlo runs with independently generated initial states. Results are shown in Figure 3.2. From the figure, we can see that $z^T(t)z(t)/n$ reduces gradually to a low level.

Figure 3.1 Information flow of network in the first simulation example. Circles labeled from *V*1 to *V*10 represent the nodes in the network. The arrow on the line indicates the direction of communication.

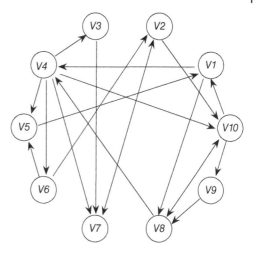

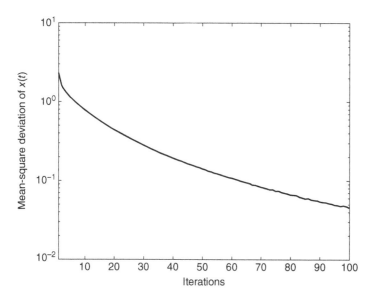

Figure 3.2 Mean-square deviations of $x(t)$ in log scales vs iteration numbers.

3.4.2 Switching and Directed Graphs

The second example studies the directed switching graph case. There are four nodes in a network with four different topologies switching in order in every 20 iterations. The information flow of the four different topologies is shown in Figure 3.3. Each node uses the consensus protocol (3.1) to update the state value. Given $\gamma = 1.6$, we design the weighting matrix for each topology.

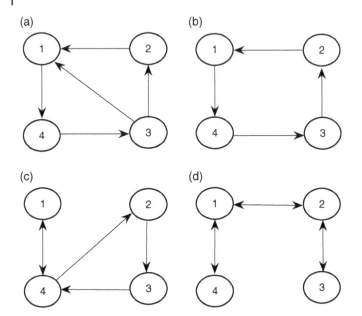

Figure 3.3 Four switching topologies in the second simulation example. The topology starts from graph (a) at time $t = 0$ and switches to the next one in every 20 iterations in the order of (a)→(b)→(c)→(d)→(a)...

We compare the result with the case using maximum-degree weights. The initial values of each node are generated according to the Gaussian distribution with zero mean and unit variance. We measure the performance by $z^T(t)z(t)/n$. Results are averaged over 500 Monte Carlo runs with independently generated initial states. In the simulation, we set $\eta = 2$ and $b = 2$ for the quantizer. Figure 3.4 shows the empirical mean-square deviation of $x(t)$ v.s. the number of iterations for the case with maximum-degree weights and that designed with our method. From Figure 3.4, we can see the mean-square deviation of our method is much lower than the case using maximum-degree weights.

3.4.3 Fixed and Directed Graphs

As mentioned before, most studies on average consensus with quantized communication assumes that the topology is undirected and fixed. To facilitate comparison, we consider here a wireless sensor network with an undirected and fixed topology. The network is generated by using a random geometric graph model [2, 12], in which nodes are uniformly distributed in a two-dimensional unit area and nodes with distance less than $\sqrt{\ln n/n}$ (n is the number of sensor nodes) are thought to be neighbors. In the simulation, we choose $n = 25$. The constructed graph in this simulation is shown in Figure 3.5. The quantization parameters are chosen as $\eta = 2, b = 4$, respectively. In each run, the state value of each node is initialized according to a Gaussian distribution with zero mean

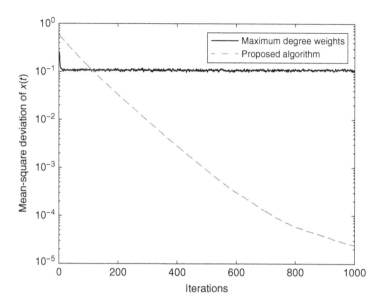

Figure 3.4 Mean-square deviations of $x(t)$ in log scales vs iteration numbers.

and unit variance and results are averaged over 500 Monte Carlo runs with independently generated initial states. We use $\gamma = 1$ to design the weighting matrix W, which is a 25×25 symmetric stochastic matrix. We compare our method with the maximum-degree weights method used in Ref. [12], the so-called LMSC method proposed in Ref. [6], and the so-called modified LMSC method with different parameters and use the mean-square deviations from the initial average as the measurement. As to the LMSC method, the weighting matrix is obtained by gradient descending search given by equation (19) in Ref. [6]. In Figure 3.6, we can see a clear advantage of using the weighting matrix obtained by the proposed algorithm over the LMSC Method and the maximum-degree weights method. After adequate iterations, the disagreement from consensus of the proposed algorithm is less than them. Note that LMSC method optimizes the weighting matrix based on the least-square deviation criterion under topological constraints. However, due to the additive noise of the LMSC method does not properly model the quantization error, the performance of the LMSC method is not as good as our method in the simulation.

3.4.4 Performance Comparison

In Ref. [6], the LMSC method is formulated as the following consensus problem with additive noises:

$$x(t + 1) = Wx(t) + \omega(t). \tag{3.51}$$

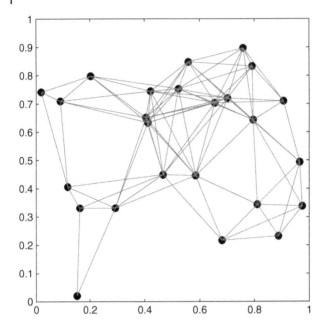

Figure 3.5 The random geometric graph used in the simulation. The number of nodes is $n = 25$. Communication radius is $\sqrt{\ln n/n}$.

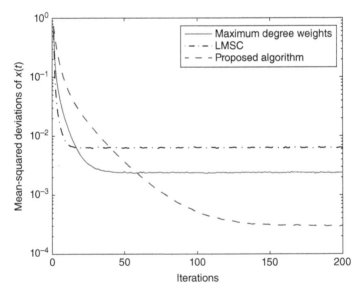

Figure 3.6 Mean-square deviations of $x(t)$ from the initial average in log scales vs iteration numbers.

Here, $\omega(t)$ is an independent and identically distributed additive noise with zero mean and constant variance. For the convenience of comparison, the consensus problem with a probabilistic quantization can be written by substituting $L = I - W$ into the equation (3.4) as follows:

$$x(t+1) = Wx(t) - v(t) + Wv(t). \tag{3.52}$$

Here, $v(t)$ is an independent and identically distributed random variable with zero mean and finite variance. Since the quantization error is a white noise, the LMSC method proposed in Ref. [6] can be extended to handle quantized communications. We compare the performances of our method with the extended LMSC method. Since we do not see any publication about the application of LMSC method in the quantized communication case, we show the detail of the extension in the following text.

Above all, the modified LMSC method employs the following assumptions:

1. The graph is undirected. That is, the weighting matrix satisfies that $W = W^T$.
2. The topology is fixed.

Note that the assumption of undirected graph is needed to guarantee the convexity of the problem. The assumption of fixed topology is required to ensure the existence of the steady-state mean-square deviation.

The problem is formulated as follows:

$$\min \qquad \delta_{ss}(W)$$
$$\text{s.t. } W = W^T, W\mathbf{1} = \mathbf{1}, W \in \mathbb{S}, \left\| W - \frac{\mathbf{1}\mathbf{1}^T}{n} \right\| < 1. \tag{3.53}$$

Here, n is the number of nodes in the graph, and δ_{ss} is the steady-state mean-square deviation. $\left\| W - \frac{\mathbf{1}\mathbf{1}^T}{n} \right\| < 1$ is the constraint for convergence and $W \in \mathbb{S}$ is the topology constraint. Note that $\delta_{ss}(W)$ has the following expression:

$$\delta_{ss} = \frac{1}{n}\text{trace}\left(\Sigma_{ss}\right). \tag{3.54}$$

Here, $\Sigma_{ss} = \lim_{t\to\infty} \Sigma_z(t)$ denotes the steady-state variance of $z(t)$ and $\Sigma_z(t)$ denotes the variance of $z(t)$ in (3.5). That is, $\Sigma_z(t) = E(z(t)z^T(t))$. As proved in equation (3.41), the objective function in (3.53) has the following form:

$$\delta_{ss}(W) = \frac{\sigma^2}{n}\sum_{i=2}^{n}\frac{1-\lambda_i}{1+\lambda_i}, \tag{3.55}$$

where λ_i denotes the ith largest eigenvalue of W and σ^2 is the quantization error variance of each node. With the expression (3.55), we are able to prove the convexity of the problem (3.53), by using the theory of convex spectral

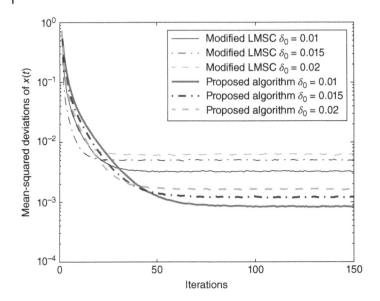

Figure 3.7 Comparisons of the proposed method and the modified LMSC method. x-axis represents the number of iterations while y-axis represent the mean-square deviations of $x(t)$ in log scales.

functions [26]. To solve this optimization problem, we can write the following convex optimization problem after some trivial operations:

$$\min g(W) = \frac{2}{n}\text{trace}\left(I + W - \frac{\mathbf{1}\mathbf{1}^T}{n}\right)^{-1}$$
$$\text{s.t.} \qquad -I - \frac{\mathbf{1}\mathbf{1}^T}{n} + W < 0$$
$$-I + \frac{\mathbf{1}\mathbf{1}^T}{n} - W < 0$$
$$W = I - \sum_{k=1}^{m} \omega_k a_k a_k^T. \tag{3.56}$$

Here, a_k denotes the kth column of the incidence matrix of the graph, ω_k denotes the weight of the kth edge, and m denotes the total amount of edges.

The modified LMSC method solves the problem (3.56) numerically to find the optimal weighting matrix. In implementation, as often done in semidefinite programming, we use the interior point method to relax problem (3.56) into an unconstrained optimization problem with the following objective function:

$$\min f(W) = \frac{2}{n}\text{trace}\left(I + W - \frac{\mathbf{1}\mathbf{1}^T}{n}\right)^{-1}$$
$$- \mu_0 \ln\left(\det\left(I - \frac{\mathbf{1}\mathbf{1}^T}{n} + W\right)\right) - \mu_0 \ln\left(\det\left(I + \frac{\mathbf{1}\mathbf{1}^T}{n} - W\right)\right).$$
$$\tag{3.57}$$

Here, μ_0 is a positive constant and $\det(\cdot)$ denotes the determinant of a square matrix. By searching in the negative gradient direction, we can find the optimal weighting matrix to problem (3.57), which is a proper approximation, by choosing a small enough μ_0, to the original problem (3.53). As observed from equation (3.55), a weighting matrix with eigenvalues $\lambda_2, \lambda_3, ..., \lambda_n$ very close to but smaller than 1, always satisfies the constraints in (3.56) and can have the objective function arbitrarily close to zero. In this way, the optimal solution of W will be infinitely close to the identity matrix. However, such a W reduces the information exchange between neighbor nodes and needs infinitely long time for convergence. In other words, the ideal solution obtained by the modified LMSC method is infeasible.

To make the modified LMSC method feasible, we set up an *allowed steady-state mean-square deviation*, δ_0, as the design criterion and run the modified LMSC method until this criterion is satisfied. In the following, we use both our method and the modified LMSC method to design weighting matrices for $\delta_0 = 0.01$, $\delta_0 = 0.015$ and $\delta_0 = 0.02$, respectively, and compare the results. As shown in Figure 3.7, the mean-square deviation of our method is lower than that of the modified LMSC method when the algorithm converges. It should be stressed that the modified LMSC method can only deal with fixed and undirected graphs. In contrast, by defining the performance index from the L_2 gain perspective, the case with switching topology and the case on directed graph can be included into our framework as stated in Theorems 3.1 and 3.5, and as depicted in the simulation examples shown in Sections 3.4.1 and 3.4.2.

3.5 Conclusion

In the chapter, average consensus on a directed graph with quantized communication was studied. We examined this problem for both fixed and switching topologies. A robust weighting matrix design was proposed to reduce the effect of the quantization error. Theoretical analysis led to a set of LMIs that can be used to solve the weighting matrices. For fixed topology with probabilistic quantization, the mean-square deviation was proved to converge and its upper bound was derived. Simulation results were given to show the effectiveness of the method.

Notes

The results in this chapter were first published in Ref. [27].

Acknowledgment is given to ©2012 John Wiley and Sons. Reprinted, with permission, from Li, S., Guo, Y., Fang, J., and Li, H. (2013) Average consensus with weighting matrix design for quantized communication on directed switching graphs. International Journal of Adaptive Control and Signal Processing, 27 (6), 519–540.

References

1 Olfati-Saber, R. (2005) Distributed Kalman filter with embedded consensus filters, in *44th IEEE Conference on Decision and Control and 2005 European Control Conference*, pp. 8179–8184, doi:10.1109/CDC.2005.1583486.

2 Xiao, L., Boyd, S., and Lall, S. (2005) A scheme for robust distributed sensor fusion based on average consensus, in *Fourth International Symposium on Information Processing in Sensor Networks*, pp. 63–70.

3 Boyd, S., Ghosh, A., Prabhakar, B., and Shah, D. (2006) Randomized gossip algorithms. *IEEE Transactions on Information Theory*, **52**, 2508–2530.

4 Fax, J. and Murray, R. (2004) Information flow and cooperative control of vehicle formations. *IEEE Transactions on Automatic Control*, **49** (9), 1465–1476.

5 Olfati-Saber, R. and Murray, R. (2004) Consensus problems in networks of agents with switching topology and time-delays. *IEEE Transactions on Automatic Control*, **49** (9), 1520–1533.

6 Xiao, L., Boyd, S., and Kim, S.J. (2007) Distributed average consensus with least-mean-square deviation. *Journal of Parallel and Distributed Computing*, **67** (1), 33–46.

7 Kar, S. and Moura, J. (2008) Distributed average consensus in sensor networks with quantized inter-sensor communication, in *IEEE International Conference on Acoustics, Speech and Signal Processing*, Las Vegas, NV, pp. 2281–2284.

8 Kashyap, A. and Srikant, R. (2007) Quantized consensus. *Automatica*, **43**, 1192–1203.

9 Cai, K. and Ishii, H. (2011) Quantized consensus and averaging on gossip digraphs. *IEEE Transactions on Automatic Control*, **56** (9), 2087–2100.

10 Li, T. and Xie, L. (2011) Distributed consensus over digital networks with limited bandwidth and time-varying topologies. *Automatica*, **47**, 2006–2015.

11 Ceragioli, F., Persis, C., and Frasca, P. (2011) Discontinuities and hysteresis in quantized average consensus. *Automatica*, **47** (9), 1916–1928.

12 Frasca, P., Carli, R., Fagnani, F., and Zampieri, S. (2009) Average consensus on networks with quantized communication. *International Journal of Robust and Nonlinear Control*, **19** (16), 1787–1816.

13 Fang, J. and Li, H. (2010) Distributed consensus with quantized data via sequence averaging. *IEEE Transactions on Signal Processing*, **58** (2), 944–948.

14 Lavaei, J. and Murray, R. (2009) On quantized consensus by means of gossip algorithm-part I: Convergence proof, in *American Control Conference*, pp. 394–401, doi:10.1109/ACC.2009.5160485.

15 Lavaei, J. and Murray, R. (2009) On quantized consensus by means of gossip algorithm-part II: Convergence time, in *American Control Conference*, pp. 394–401, doi:10.1109/ACC.2009.5160485.

16 Carli, R., Fagnani, F., Frasca, P., and Zampieri, S. (2010) Gossip consensus algorithms via quantized communication. *Automatica*, **46** (1), 70–80.

17 Ren, W., Beard, R., and Atkins, E. (2005) A survey of consensus problems in multi-agent coordination, in *2005 American Control Conference*, vol. 3, Portland, OR, pp. 1859–1864.

18 Lin, P., Jia, Y., and Li, L. (2008) Distributed robust H_∞ consensus control in directed networks of agents with time-delay. *Systems & Control Letters*, **57** (8), 643–653.

19 Boyd, S., Diaconis, P., and Xiao, L. (2004) Fastest mixing Markov chain on a graph. *SIAM Review*, **46** (4), 667–689.

20 Xiao, L. and Boyd, S. (2004) Fast linear iterations for distributed averaging. *Systems and Control Letters*, **53**, 65–78.

21 Jakovetic, D., Xavier, J., and Moura, J. (2010) Weight optimization for consensus algorithms with correlated switching topology. *IEEE Transactions on Signal Processing*, **58** (7), 3788–3801.

22 Boyd, S., El Ghaoui, L., Feron, E., and Balakrishnan, V. (1994) *Linear matrix inequalities in system and control theory*, SIAM Studies in Applied Mathematics, Society for Industrial and Applied Mathematics (SIAM), Philadelphia, PA.

23 Meyer, C. (2000) *Matrix Analysis and Applied Linear Algebra*, SIAM: Society for Industrial and Applied Mathematics, Philadelphia.

24 Cvetkovic, D., Doob, M., Sachs, H., Cvetkovi, A., and Horstlistprice, M. (1998) *Spectra of Graphs: Theory and Applications*, 3rd Revised and Enlarged Edition, Vch Verlagsgesellschaft Mbh, New York.

25 Liberzon, D. (2003) *Switching in Systems and Control*, 1st edn., Birkhauser, Boston.

26 Borwein, J. and Lewis, A. (2000) *Convex Analysis and Nonlinear Optimization, Theory and Examples*, Canadian Mathematical Society Books in Mathematics, Springer-Verlag, New York.

27 Li, S., Guo, Y., Fang, J., and Li, H. (2013) Average consensus with weighting matrix design for quantized communication on directed switching graphs. *International Journal of Adaptive Control and Signal Processing*, **27** (6), 519–540.

4

Weighted Average Consensus for Cooperative Spectrum Sensing

4.1 Introduction

Cognitive radio (CR) [1] aims to improve the spectrum utilization by allowing unlicensed secondary user (SU) to operate in the "white spaces" of the licensed spectrum bands without interfering the licensed primary user (PU). Revealing a future communication paradigm with dramatically enhanced spectrum efficiency, CR network is also referred as the neXt generation (XG) or dynamic spectrum access (DSA) network [2].

One of the fundamental techniques in CR is spectrum sensing, which enables the SUs users to detect the presence of a PU in the spectrum; see Refs. [3], [4] and the references therein. The main challenge of spectrum sensing is the receiver uncertainty problem [1] such as practical multipath fading and shadowing, which compromise the detecting performance significantly. Recent research progress shows cooperative spectrum sensing [5] is a promising methodology to improve the spectrum sensing performance under shadowing, fading, and time-varying wireless channels.

Related Work in Cooperative Spectrum Sensing

The main advantage of cooperative spectrum sensing is to enhance the sensing performance by exploiting the observation diversity of spatially located SUs [5]. By cooperation, CR users can share their sensing information to make a combined decision which is more accurate than individual decisions. Cooperative sensing usually contains two stages: sensing and fusion. In the sensing stage, each SU makes the measurement using appropriate detecting techniques. Among all types of detectors, energy detector is widely applied because it requires lower design complexity and no prior knowledge of PUs as compared to other techniques such as matched filter detection or cyclostationary detection [6]. In the fusion stage, the SU network cooperatively combines the detecting statistics throughout the network and the final decision is made using global information. Among the fusion techniques, different measurement combining methods have been considered including hard bit combining [7], soft gain combining [8], to name a few.

The key element of cooperative sensing is the cooperation scheme, which decides the SU network structure and the detecting performance. Centralized cooperative sensing and relay-assisted cooperative sensing are two major schemes in the literature [5]. Centralized cooperative sensing [9] lets all SUs report their measurement information to a centralized fusion center; then a global decision is made at the fusion center according to certain measurement combining methods. Relay-assisted cooperative sensing [5], [10] is a multi-hop cooperation scheme that makes use of the strong sensing channels and strong reporting channels among the SU network in order to improve the overall performance. Relay-assisted sensing can be either centralized with a fusion center, or distributed without a fusion center. Centralized cooperative spectrum sensing requires that the entire received data be gathered at one place, which may be difficult due to communication constraints [11]. The multi-hop communication of the relay-assisted sensing may bring extra power cost than one-hop communication, since all SUs' sensing data need to be relayed from the network nodes to the fusion center or detection node. In addition, the multi-hop communication paths may degrade the sensing data quality and affect the detection performance significantly compared to one-hop communication scenarios. Other factors such as communication channel selection schemes and sensing data coding schemes also need to be considered [12] in the relay assisted cooperative sensing to overcome the disadvantage of the multi-hop paths.

Distributed cooperative sensing first appears in Ref. [7] with broadcasting schemes. After measurement, each SU broadcasts its own decision to all SU nodes in the network, and the final decision is decided by OR rule. Very recently, bioinspired consensus scheme is introduced to spectrum sensing in Refs. [13], [14] for distributed measurement fusion and soft combining. Consensus-based spectrum sensing is a biologically inspired approach learned from swarming behaviors of fish schools and bird fleets. The consensus-based cooperation features self-organizable and scalable network structure and only needs one-hop communication among local neighbors. Recent research work [15] applies belief propagation to distributed spectrum sensing [15], which advances the sensing stage for heterogenous radio environment.

The fusion scheme of the sensing data from the SU network also contribute to the detection performance. There are hard bit combining such as OR rule combining and soft combining including equal gain combining and weighted gain combining. Hard bit combining adopts the decision bit from each SU to achieve global detection, which is less effective compared to soft combing schemes taking average of the statistics from all the SUs. Generally speaking, equal gain combing is to compute the average of the measured statistics of the SU network, while weighted gain combining computes the weighted average considering the measurement channel conditions. Therefore, weighted combining offers better detection performance under various channel conditions such as fading and shadowing.

The future CR networks will most likely consist of smartphones, tablets, and laptops moving with the swarming behaviors of people. Therefore, consensus-based spectrum sensing reveals great potential for future development of distributed CR networks. However, the existing consensus-based fusion algorithms [14, 16] only ensure equal gain combining of local measurements, which is incomparable with centralized *weighted* combining approaches [8]. To make the distributed consensus-based spectrum sensing more robust to practical channel conditions and link failures, we need to develop new distributed *weighted* fusion algorithms which are missing in the current literature.

Related Work in Average Consensus

The consensus algorithm was studied in Ref. [17] for modeling decentralized decision making and parallel computing. The main benefit of consensus is ensuring each node to hold the global average of the initial values throughout the network using local communication between one-hop neighboring nodes. Two decades later, consensus algorithm is introduced to multiagent systems [18, 19]. In Ref. [18], Jadbabaie et al. analyze the convergence conditions of a biologically rooted discrete-time consensus model, but the convergence value is not specified. Olfati-Saber and Murray give the conditions for average consensus convergence of continuous time consensus model in Ref. [19]. Since the average consensus problem has strong impact on distributed networked systems, it increasingly attracts research attention on decentralized estimation [20], filtering [21], and detection [22], etc. For signal processing applications, communication constraints and the convergence rate become crucial for performance improvement. Typical problems include communication topology design and optimization [23], convergence rate analysis, and optimization [24]. Interested readers are referred to the review papers [25, 26] for the complete history of consensus algorithm development.

Compared to the extensively studied average consensus, much less research attention is paid to *weighted* average consensus. As stated in Ref. [25], weighted average consensus algorithm is modeled by asymmetric matrices which makes the mathematical tools for average consensus algorithm inapplicable, and it is difficult to predict the convergence value on dynamic communication channels. However, weighted average consensus algorithm in the fusion process of spectrum sensing can achieve *weighted* gain combining without a fusion center, which advances the consensus-based spectrum sensing significantly. Therefore, it is important to develop solid theoretical analysis of weighted average consensus algorithms on dynamic communication topologies.

Main Contribution of This Chapter

In this chapter, we propose a distributed cooperative spectrum sensing scheme based on weighted average consensus algorithm. We adopt weighted average consensus algorithm for measurement fusion of the distributed cooperative spectrum sensing. Weighted average consensus-based fusion allows each SU to choose a weight according to the measurement condition, and the global fused statistic is a soft weighted combining reflecting the measurement quality without centralized fusion center. Compared to the existing average consensus based approaches [13, 14], the proposed method offers better detection performance, and achieves comparable performance with the centralized weighted combining method [27, 28]. With rigorous theoretic proof, the proposed method is robust with respect to dynamic communication channel conditions. The weighted average consensus-based spectrum sensing provides a generic distributed weighted combining and applies to both additive white Gaussian noise (AWGN) channels and Rayleigh fading channels.

The main contribution of this chapter has two-folds. First, we provide formal convergence analysis of the weighted average consensus under fixed and dynamic communication channels, which advances the theoretical development of consensus algorithms and encompasses average consensus as a special case. In particular, we rigorously prove that temporary communication link failures do not affect the convergence of the weighted average consensus under the jointly connected condition. Second, we apply weighted consensus-based distributed weighted soft combining method in cooperative spectrum sensing, and provide a formal treatment of the distributed sensing algorithm in this chapter. We obtain closed-form optimal weight design in the distributed weighted combining scheme for the generic additive Gaussian channel approximation, and estimate the convergence rate of the consensus iteration under the assumption that each communication link has an independent probability to fail. We characterize the upper bound of the iteration number of the ϵ-convergence, which indicates all SUs are ϵ close to the final convergence value in the probability sense. Simulation results show significant improvement of the sensing performance compared to existing consensus-based approaches, and the performance of the distributed weighted design is comparable to centralized weight combining schemes.

Organization: The rest of the chapter is organized as follows. In Section 4.2, the spectrum sensing model using energy detector is introduced as preliminaries. Section 4.3 presents the main results on weighted average fusion algorithm and weight design choices for AWGN channels and Rayleigh fading channels. Section 4.4 illustrates the proofs for the weighted average consensus convergence for both fixed and dynamic communication channels. Section 4.5 presents the simulation results, and Section 4.6 concludes the chapter.

4.2 Problem Statement

A scenario of cooperative spectrum sensing is illustrated in Figure 4.1. The TV signals shown in the figure is considered as PU signals, and three CR nodes, denoted as CR 1, CR 2, and CR 3 in the figure, are SUs. The scenario shows that environmental diversity differentiates the signal quality of the CR nodes, that

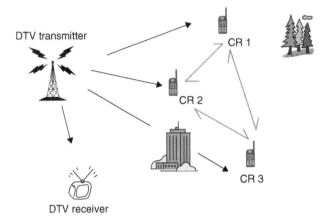

DTV transmitter

CR 1

CR 2

CR 3

DTV receiver

Figure 4.1 A scenario of cooperative spectrum sensing.

is, CR 1 and CR 2 have better locations to make the detection measurements, while CR 3 is blocked by a building; thus, its detection performance is degraded by fading effects, for example. If we consider the TV RF signal as a virtual "light" source, CR 1 and CR 2 have better vision to "see" the source, while CR 3 is located in the shadow thus has worse "vision." Since the CR nodes have different measurement qualities, combining the receiving signals with equal gains using *average* consensus is apparently not optimal. Indeed, the symmetric structure of average consensus does not match the asymmetry of network location diversities caused by the environment. In this case, shifting *average* consensus to *weighted* average consensus to match the asymmetry of environment will improve the overall detection performance.

Consensus-based cooperative spectrum sensing adopts a two-stage detection scheme: sensing and fusion. For each detection time slot, in the sensing stage, each SU makes measurement of the PU signal using energy detectors. In the fusion stage, each SU communicates with its neighbors using the consensus algorithm to fuse the measurement until convergence. After the fusion convergence, each SU holds the global combined statistic and makes the decision with a global point of view. In this section, we introduce the fundamentals of energy detector and existing centralized weighted combining model of SU networks using energy detection.

To detect the existence of an unknown signal, the energy detector [29] is the first choice, since it requires lower design complexity and no prior knowledge of PUs. For each detector, the received signal $y_i(t)$ is modeled as

$$y_i(t) = \begin{cases} n_i(t), & \mathcal{H}_0 \\ h_i s(t) + n_i(t), & \mathcal{H}_1, \end{cases} \tag{4.1}$$

where $\mathcal{H}_0$ represents the hypothesis without PU's signal; $\mathcal{H}_1$ represents the hypothesis with PU's signal; h_i represents gain of the channel, which is assumed to be constant for each sampling; $s(t)$ is the signal from PU and; $n_i(t)$ is the AWGN, that is, $n_i(t) \sim \mathcal{N}(0, \sigma_i^2)$. We call $n_i(t)$ the sensing noises and collect their variances into a vector

$$\sigma = [\sigma_1^2, \sigma_2^2, \dots, \sigma_n^2]^T. \tag{4.2}$$

Without loss of generality, $s(t)$ and $\{n_i(t)\}$ are assumed to be independent of each other.

According to Urkowitz [30], each SU calculates a summary statistic Y_i over a detection interval T with m samples, that is,

$$Y_i = \sum_{t=0}^{m-1} |y_i(t)|^2 \quad i \in \mathcal{I}, \tag{4.3}$$

where m is determined by two times of the time-bandwidth product TW, that is, over the sampling duration T, $m = 2TW$ samples are collected with $(1/2W)$ apart for any two consecutive samples.

Under AWGN measurement channels, the test statistic of the ith SU using energy detection is given by (4.3). Since Y_i is the sum of the squares of m Gaussian random variables, it can be shown that Y_i/σ_i^2 follows a central chi-square χ^2 distribution with m degrees of freedom if $\mathcal{H}_0$ is true; otherwise, it would follow a noncentral χ'^2 distribution with m degrees of freedom and parameter η_i. That is,

$$\frac{Y_i}{\sigma_i^2} = \begin{cases} \chi_m^2, & \mathcal{H}_0 \\ \chi'^2_m(\eta_i), & \mathcal{H}_1, \end{cases} \tag{4.4}$$

where

$$\eta_i = \frac{E_s |h_i|^2}{\sigma_i^2} \tag{4.5}$$

is the local SNR at the ith SU and the quantity

$$E_s = \sum_{t=0}^{m-1} |s(t)|^2 \tag{4.6}$$

represents the transmitted signal energy over a sequence of m samples during each detection interval. Note that the so-defined local SNR is m times the average SNR at the output of the energy detector, which should be equal to $E_s |h_i|^2 / m\sigma_i^2$. For convenience, we put all η_i into the following vector:

$$\eta = [\eta_1, \eta_2, \dots, \eta_n]^T. \tag{4.7}$$

According to the central limit theorem, if the number of samples m is large enough (e.g., ≥ 10 in practice), the test statistics Y_i are asymptotically normally distributed with mean

$$E(Y_i) = \begin{cases} m\sigma_i^2 & \mathcal{H}_0 \\ (m + \eta_i)\sigma_i^2 & \mathcal{H}_1 \end{cases} \tag{4.8}$$

and variance

$$\text{Var}(Y_i) = \begin{cases} 2m\sigma_i^4 & \mathcal{H}_0 \\ 2(m + 2\eta_i)\sigma_i^4 & \mathcal{H}_1 \end{cases}. \tag{4.9}$$

Gaussian distribution approximation will facilitate the optimal weight design of the soft weighted combining.

Existing centralized cooperative spectrum sensing schemes combines the measurements of the SUs at a fusion center as [27, 28]

$$Y_g = \sum_{i=1}^{n} \omega_i Y_i = \omega^T Y, \tag{4.10}$$

where $\omega = [\omega_1, \omega_2, \dots, \omega_n]^T$, $\omega_i \geq 0$ is the weighting ratio, $Y = [Y_1, Y_2, \dots, Y_n]^T$ is the measurement of the CR network.

Assume the reporting channel is noise-free and all $\{Y_i\}$ are assumed to be normal random variables; Y_g is also normally distributed and has mean

$$\overline{Y}_g = EY_g = \begin{cases} m\sigma^T\omega & \mathcal{H}_0 \\ (m\sigma + E_s g)^T \omega & \mathcal{H}_1 \end{cases}, \tag{4.11}$$

where

$$\sigma = [\sigma_1^2, \sigma_2^2, \dots, \sigma_n^2]^T, \tag{4.12}$$

$$g = [|h_1|^2, |h_2|^2, \dots, |h_3|^2]^T, \tag{4.13}$$

and the variances under different hypotheses are, respectively, given by

$$\text{Var}(Y_g | \mathcal{H}_l) = E(Y_g - \overline{Y}_g)^2$$
$$= \omega^T E\left[(Y - \overline{Y}_{\mathcal{H}_k})(Y - \overline{Y}_{\mathcal{H}_k})^T | \mathcal{H}_k\right]\omega \tag{4.14}$$

where $\overline{Y} = EY$ and $l \in \{0, 1\}$, specifically, we have

$$\text{Var}(Y_g | \mathcal{H}_l) = \begin{cases} \omega^T \Sigma_{\mathcal{H}_0} \omega & \text{under } \mathcal{H}_0 \\ \omega^T \Sigma_{\mathcal{H}_1} \omega & \text{under } \mathcal{H}_1 \end{cases} \tag{4.15}$$

Here,

$$\Sigma_{\mathcal{H}_0} = 2m\text{diag}^2(\sigma), \tag{4.16}$$

$$\Sigma_{\mathcal{H}_1} = 2m\text{diag}^2(\sigma) + 4\text{diag}(\eta)\text{diag}^2(\sigma). \tag{4.17}$$

With a test threshold λ, we have

$$Y_g \underset{\mathcal{H}_0}{\overset{\mathcal{H}_1}{\gtrless}} \lambda, \tag{4.18}$$

and the performance of the proposed cooperative spectrum detection scheme can be evaluated as

$$P_f = Q\left[\frac{\lambda - N\sigma^T\omega}{\sqrt{\omega^T\Sigma_{\mathcal{H}_0}\omega}}\right] \tag{4.19}$$

and

$$P_d = Q\left[\frac{Q^{-1}(P_f)\sqrt{\omega^T \Sigma_{H_0} \omega} - E_s g^T \omega}{\sqrt{\omega^T \Sigma_{H_1} \omega}}\right]. \tag{4.20}$$

Given a fixed false alarm P_f, maximizing P_d in Equation (4.20) will yield the optimal weights ω, see Ref. [28], where an optimal solution based on modified deflection coefficient are discussed on centralized soft combining.

4.3 Cooperative Spectrum Sensing Using Weighted Average Consensus

In this section, we present our new consensus-based distributed scheme to achieve the weighted measurement combining through local interactions among SUs, instead of processing the measurements in a centralized fusion center.

4.3.1 Weighted Average Consensus Algorithm

Weighted average consensus algorithm is the key for the distributed weighted combining, as introduced in this section. After the sensing stage, we assume that each SU obtains the measurement of the PU's signal energy. Following the notation of consensus [25], we denote the value of the ith agent's measurement Y_i as x_i, and the proposed weighted average consensus-based combining scheme is

$$x_i(k+1) = x_i(k) + \frac{\alpha}{\delta_i} \sum_{j \in N_i(k)} (x_j(k) - x_i(k)), i \in I, \tag{4.21}$$

where α is the iteration stepsize satisfying the maximum node degree constraint [25], $N_i(k)$ denotes neighboring node of the ith SU at time step k, and $\delta_i \geq 1$ is the weighting ratio.

If the communication topologies formed by the SU network are jointly connected, all the SUs' decision statistics will reach consensus. The final convergence value is as follows:

$$x_i(k) \to x^* = \frac{\sum_{i=1}^n \delta_i x_i(0)}{\sum_{i=1}^n \delta_i} \text{ as } k \to \infty, \forall i \in I. \tag{4.22}$$

The rigorous proof of the convergence is given in Section 4.4. By comparing the decision value x^* with a predefined threshold λ, every SU locally obtains the global decision as follows:

$$\text{Decision } H = \begin{cases} \mathcal{H}_1, & x^* > \lambda \\ \mathcal{H}_0, & \text{Otherwise.} \end{cases} \tag{4.23}$$

Remark 4.1   An important feature of the algorithm is the choice of the weighting factor δ_i. If the ith SU has better measurement channel condition, it sets a larger δ_i, which makes the iteration (4.21) rely less on the local information exchange. On the contrary, an SU with poor measurement channel sets a smaller δ_i and relies more on the information from the network in order to improve the overall performance. Therefore, the statistics value across the whole SU network will be dominated by the SUs with better measurements. For example, setting the weight δ_i in Equation (4.21) as $\delta_i = \overline{\gamma}_i$, $\forall i \in \mathcal{I}$, $\overline{\gamma}_i$ is the estimated average SNR. The final consensus value will be $x_i(k) \to x^* = \frac{\sum_{i=1}^{n} \overline{\gamma}_i x_i(0)}{\sum_{i=1}^{n} \overline{\gamma}_i}$ [31].

Remark 4.2   The average consensus-based combining approach [14, 16, 22] is a special case in our proposed algorithm when $\delta_i = 1$, $\forall i \in \mathcal{I}$. Compared to the centralized soft combining in Equation (4.10), the distributed consensus iteration in Equation (4.21) achieves an equivalent combining as

$$\omega_i = \frac{\delta_i}{\sum_{i=1}^{N} \delta_i}. \tag{4.24}$$

After the convergence, the final decision statistic x^* equals the global combing Y_g, and every SU holds a weighted global decision consensus only through local information exchange.

4.3.2 Fusion Convergence Performance in Terms of Detection Probability

The distributed fusion based on (4.21) is a iterative process that completes after the convergence is reached. In this section, we characterize the convergence of the proposed fusion algorithm in terms of the detection probability. If we write the algorithm of Equation (4.21) in the compact form,

$$x(k+1) = W(k)x(k), \tag{4.25}$$

where $x = [x_1, \ldots, x_n]^T$, and $W(k)$ is the iteration transition matrix at time step k. We will prove in Section 4.4 that

$$\lim_{k \to \infty} \prod_{i=1}^{k} W(i) = \frac{1\delta^T}{\delta^T 1}, \tag{4.26}$$

where

$$\delta = [\delta_1, \delta_2, \ldots, \delta_n]^T, \tag{4.27}$$

where δ_i is the weighting ratio in the algorithm (4.25).

If we assume that $x_i(0)$ is the energy detector output under AWGN measurement channels and follows a normal distribution as discussed in Section 4.2, we have

$$x_i(k) = [W]_i x(0), \tag{4.28}$$

where $[W]_i$ denotes the ith row of the matrix $\prod_{i=1}^{k} W(k)$. Therefore, $x_i(k)$ is a weighted average of Gaussian-distributed random variables, which is also Gaussian distributed; that is,

$$\mathcal{H}_0 : x(k)_i \sim \mathcal{N} \left\{ m \sum_{j=1}^{n} w_{ij}\sigma_i^2, \sqrt{2m \sum_{j=1}^{n} w_{ij}^2 \sigma_i^4} \right\}$$

$$\mathcal{H}_1 : x(k)_i \sim \mathcal{N} \left\{ \sum_{j=1}^{n} w_{ij}(m + \eta_i)\sigma_i^2, \sqrt{\sum_{j=1}^{n} w_{ij}^2 2(m + 2\eta_i)\sigma_i^4} \right\},$$

where w_{ij} is the element of matrix $\prod_{i=1}^{k} W(k)$ at the ith row and jth column. Thus, the probability of detection at the ith SU at time k is given by

$$P_f(k)_i = Q(\lambda; \mu_0, \nu_0) \tag{4.29}$$
$$P_d(k)_i = Q(\lambda; \mu_1, \nu_1), \tag{4.30}$$

where $Q(\cdot)$ is the complementary cumulative distribution function of Gaussian variable, λ is the decision threshold, and

$$\{\mu_0, \nu_0\} = \left\{ m \sum_{j=1}^{n} w_{ij}\sigma_i^2, \sqrt{2m \sum_{j=1}^{n} w_{ij}^2 \sigma_i^4} \right\}$$

$$\{\mu_1, \nu_1\} = \left\{ \sum_{j=1}^{n} w_{ij}(m + \eta_i)\sigma_i^2, \sqrt{\sum_{j=1}^{n} w_{ij}^2 2(m + 2\eta_i)\sigma_i^4} \right\}.$$

From Equation (4.26), we have

$$\lim_{k \to \infty} w_{ij} = \delta_j, \tag{4.31}$$

where δ_j is the weighting ratio in the algorithm (4.21).

Practically, it's unnecessary to process the algorithm for the infinite iteration. We can use Equation (4.30) with respect to the time step k as an evaluation for the transient performance in finite steps of the consensus-based cooperative spectrum sensing schemes.

4.3.3 Optimal Weight Design under AWGN Measurement Channels

In this section, we describe the distributed optimal weight design based on the proposed weighted average consensus algorithm. First, we consider the combined global statistic in Equation (4.10) and obtain the optimized centralized weights ω_{oi}. Then, using Equation (4.24), we obtain the distributed optimal weights δ_{oi}.

Given a false-alarm constraint P_f, the optimal weights can be obtained by maximizing P_d. Generally speaking, the closed-form solution does not exist for maximizing P_d in Equation (4.20). To give an optimal weight design, we consider maximizing the deflection coefficient [32] for a near-optimal weight design. Specifically, the centralized optimal solution is

$$\omega_{oi} = \frac{\frac{\eta_i}{\sigma_i^2}}{\sum_{i=1}^{n} \frac{\eta_i}{\sigma_i^2}}, \tag{4.32}$$

where ω_{oi} denotes the optimal value of ω_i, η_i is the local SNR defined in Equation (4.5), and σ_i^2 is the variance of the Gaussian noise in the measurement channel. Comparing Equation (4.24) and (4.32), we obtain the distributed weights as follows:

$$\delta_{oi} = \frac{\eta_i}{\sigma_i^2}. \tag{4.33}$$

Remark 4.3 Because the weighted average consensus ensures the linear combining, the uniformed weights should be in a linear form as in Equation (4.32). All the δ_{oi} need to be scaled or saturated to be larger than 1 without affecting the convergence of the consensus iteration under independent and identically distributed (i.i.d.) AWGN channel.

Remark 4.4 The proposed weight design in the form (4.32) or (4.33) is a suboptimal approximated solution. Similar weights design are proposed in Refs. [8, 27, 28] based on different assumptions and different approximation methods, such as maximizing modified deflection coefficient and likelihood ratio test. The point we want to emphasize is that optimal weight design can easily be incorporated in our weighted consensus algorithm (4.21) to fuse the measurement statistics for certain optimal weighted combining according to different channel conditions.

To show the optimality of the weight in Equation (4.33), we define the *deflection coefficient*, based on the cooperative spectrum sensing settings, as follows:

$$d^2(\omega) = \frac{\left[E(Y_g|\mathcal{H}_1) - E(Y_g|\mathcal{H}_0)\right]^2}{\text{Var}(Y_g|\mathcal{H}_0)} = \frac{(E_s g^T \omega)^2}{\omega^T \Sigma_{\mathcal{H}_0} \omega}. \tag{4.34}$$

Here, $E(Y_g)$ and $\text{Var}(Y_g)$ under the hypothesis $\mathcal{H}_0$ and $\mathcal{H}_1$ are defined in (4.11) and (4.14), respectively. Rewriting Equation (4.20) as

$$P_d = Q\left[\frac{Q^{-1}(P_f) - \frac{E_s g^T \omega}{\sqrt{\omega^T \Sigma_{\mathcal{H}_0} \omega}}}{\sqrt{\frac{\omega^T \Sigma_{\mathcal{H}_1} \omega}{\omega^T \Sigma_{\mathcal{H}_0} \omega}}}\right]$$

$$= Q\left[\frac{Q^{-1}(P_f) - \frac{E_s g^T \omega}{\sqrt{\omega^T \Sigma_{\mathcal{H}_0} \omega}}}{\sqrt{1 + \frac{4\omega^T \text{diag}(\eta)\text{diag}^2(\sigma)\omega}{\omega^T \Sigma_{\mathcal{H}_0} \omega}}}\right]. \tag{4.35}$$

where $Q(\cdot)$ denotes the complementary cumulative distribution function. From Equation (4.35), we can see that in low signal-to-noise ratio (SNR) channel condition when $\frac{4\omega^T \text{diag}(\eta)\text{diag}^2(\sigma)\omega}{\omega^T \Sigma_{H_0}\omega} \ll 1$, maximizing $\frac{(E_s g^T \omega)^2}{\omega^T \Sigma_{H_0}\omega}$ will yield a near-optimal weight design. We formulate the problem as follows:

$$\max_\omega d^2(\omega) \tag{4.36}$$

$$\text{st.} \quad \sum_{i=1}^N \omega_i = 1, \quad \omega_i > 0, \forall i \in \mathcal{I}.$$

Solving (4.36), we can obtain optimal distributed solution using Equation (4.24). Substituting $\omega' = \Sigma_{H_0}^{1/2}\omega$ into (4.34) yields

$$d^2(\omega) = \frac{E_s^2 \omega'^T \Sigma_{H_0}^{-T/2} gg^T \Sigma_{H_0}^{-1/2}\omega'}{\omega'^T \omega'}$$

$$\leq E_s^2 \lambda_{\max}\left(\Sigma_{H_0}^{-T/2} gg^T \Sigma_{H_0}^{-1/2}\right). \tag{4.37}$$

We can see that the matrix $\Sigma_{H_0}^{-T/2} gg^T \Sigma_{H_0}^{-1/2}$ is a rank one matrix having the nonzero eigenvalue $\lambda_{\max} = \|\Sigma_{H_0}^{-T/2} g\|_2^2$, and the associated eigenvector $\Sigma_{H_0}^{-T/2} g$. Let

$$\omega' = \Sigma_{H_0}^{-T/2} g, \tag{4.38}$$

$d^2(\omega)$ will achieve the maximum value $E_s^2 \|\Sigma_{H_0}^{-T/2} g\|_2^2$. Therefore, the uniformed optimal weight is

$$\omega_o = \frac{\Sigma_{H_0}^{-1/2}\omega'}{1^T \Sigma_{H_0}^{-1/2}\omega'} = \frac{\Sigma_{H_0}^{-1} g}{1^T \Sigma_{H_0}^{-1} g}. \tag{4.39}$$

Because Σ_{H_0} defined in Equation (4.16) is a diagonal matrix, we have

$$\omega_{oi} = \frac{\frac{|h_i|^2}{2m\sigma_i^4}}{\sum_{i=1}^n \frac{|h_i|^2}{2m\sigma_i^4}} = \frac{\frac{\eta_i}{\sigma_i^2}}{\sum_{i=1}^n \frac{\eta_i}{\sigma_i^2}}. \tag{4.40}$$

Using Equation (4.24), we can choose

$$\delta_{oi} = \frac{\eta_i}{\sigma_i^2} \tag{4.41}$$

as a distributed optimal design. Thus, the final consensus value is the near-optimal soft weighted combining.

4.3.4 Heuristic Weight Design under Rayleigh Fading Channels

We considered optimal weight design under AWGN channel in Section 4.3.3. Under different channel conditions, we shall consider different schemes for choosing the weighting parameter δ_i. In this section, we discuss the weight design under Rayleigh fading channels.

Following Refs. [29, 33], under Rayleigh fading channels, the channel gain h is random and the resulting SNR γ follows an exponential distribution. The output Y_i of the energy detector of the ith agent has the following distributions under hypothesis:

$$Y_i \sim \begin{cases} \chi_m^2, & \mathcal{H}_0 \\ \chi'^2_{m-2} + Y_e, & \mathcal{H}_1. \end{cases} \tag{4.42}$$

Here, χ_m^2 and χ'^2_{m-2} denote random quantities under central chi-square distributions with degree of freedom m and $m-2$, respectively. Y_e has an exponential distribution with parameter $2(\bar{\gamma}+1)$, $\bar{\gamma}$ represents the average SNR of the fading channels. The hypothesis $\mathcal{H}_0$ and $\mathcal{H}_1$ denote the absence and presence, respectively, of the PU.

Since the distributions under $\mathcal{H}_0$ and $\mathcal{H}_1$ hold the mean m and $m+2\bar{\gamma}$, respectively, each SU can simply estimates the average SNR from recent measurements as [34] follows:

$$\bar{\gamma}_i = \frac{1}{2l} \sum_{j=k-l}^{k} (Y_{i,j} - m). \tag{4.43}$$

Here, $Y_{i,j}$ is the jth measurement of the ith SU and l is the estimation window.

General closed form of optimal weight design under Rayleigh fading channels does not exist. We could simply adopt the weights from estimated average channel SNR (4.43). From $\mathcal{H}_1$ in (4.42), setting the weight δ_i in Equation (4.21) as $\delta_i = \bar{\gamma}_i$, $\forall i \in I$, $\bar{\gamma}_i$ is the estimated average SNR, The final consensus value will be $x_i(k) \to x^* = \frac{\sum_{i=1}^n \bar{\gamma}_i x_i(0)}{\sum_{i=1}^n \bar{\gamma}_i}$.

Remark 4.5 For the centralized weighted combining (4.10), the centralized weights ω_i can be chosen as $\omega_i = \frac{\bar{\gamma}_i}{\sum_{i=1}^m \bar{\gamma}_i}$ [34], accordingly for Rayleigh fading measurement channels.

4.4 Convergence Analysis

In this section, we rigorously prove the convergence of the consensus-based combining algorithm in Equation (4.21) under fixed and dynamic communication channel conditions. We further characterize the convergence rate assuming each communication link has a failure probability.

4.4.1 Fixed Communication Channels

Analyzing the convergence of the algorithm (4.21) under fixed communication channels will bring basic understanding of the weighted consensus algorithm and help the analysis of the algorithm under dynamic communication channels. For convenience, we rewrite the algorithm (4.21) in the following compact form:

$$x(k+1) = Wx(k). \tag{4.44}$$

Here, $x = [x_1, \ldots, x_n]^T$, and W is defined as

$$W = I - \alpha \Delta^{-1} L, \tag{4.45}$$

where $\Delta = \text{diag}\{\delta_1, \ldots, \delta_n\}$, $L \in \mathbb{R}^{n \times n}$ is the Laplacian matrix as defined in Chapter 2. The stepsize α satisfies

$$0 < \alpha < \frac{1}{d_{\max}}, i \in \mathcal{I}. \tag{4.46}$$

The convergence of Equation (4.44) depends on the convergence of the infinite matrix product

$$\lim_{k \to \infty} W^k = \frac{\mathbf{1}\delta^T}{\delta^T \mathbf{1}}, \tag{4.47}$$

where

$$\delta = [\delta_1, \delta_2, \ldots, \delta_n]^T, \tag{4.48}$$

and δ^T is the left eigenvector of W associated with the eigenvalue 1. We have Theorem 4.1.

Theorem 4.1   For the iteration process (4.21), if the stepsize α satisfies maximum node degree constraint (4.46), and the elements of matrix $\Delta = \text{diag}\{\delta_1, \ldots, \delta_n\}$ satisfy $\delta_i \geq 1, \forall i \in \mathcal{I}$, and the communication graph is fixed, then the iteration exponentially converges to

$$\lim_{k \to \infty} W^k x(0) = \frac{\sum_{i=1}^{n} \delta_i x_i(0)}{\sum_{i=1}^{n} \delta_i} \mathbf{1}. \tag{4.49}$$

That is,

$$x^* = \lim_{k \to \infty} x(k) = \frac{\sum_{i=1}^{n} \delta_i x_i(0)}{\sum_{i=1}^{n} \delta_i}. \tag{4.50}$$

Proof: The proof mainly follows from the spectral decomposition and Perron–Frobenius theorem [35]. To characterize the convergence of the algorithm (4.44), we make the *spectral decomposition* [35] of W, shown as

$$W = \lambda_1 J_1 + \lambda_2 J_2 + \cdots + \lambda_n J_n, \tag{4.51}$$

where $\lambda_i \in \mathbb{C}^1, i \in \mathcal{I}$ is the spectrum of the W, $J_i, i \in \mathcal{I}$ are the *spectral projectors* of W. J_i's satisfy the following conditions:

1. J_i is the projector onto $N(W - \lambda_i I)$ along $R(W - \lambda_i I)$, $J_i J_i = J_i$.
2. $J_i J_j = 0$ whenever $i \neq j$.
3. $J_1 + J_2 + \cdots + J_n = I$.

For the spectrum projector, we have Lemma 4.1.

Lemma 4.1 *([35])*   If x and y^T are respective right and left eigenvectors associated with a simple eigenvalue of matrix W, then the spectrum projector associated with the simple eigenvalue is

$$J = \frac{xy^T}{y^T x}. \tag{4.52}$$

Following the spectrum decomposition, we obtain

$$\lim_{k \to \infty} W^k = \lim_{k \to \infty} \left(\lambda_1^k J_1 + \lambda_2^k J_2 + \cdots + \lambda_n^k J_n \right). \tag{4.53}$$

From Equation (4.53), we can see that $\rho(W) \leq 1$ is necessary for the convergence of the algorithm (4.44), and the eigenvalue $\lambda_i = 1$ will decide the final convergence value. Note that the spectral decomposition and projection are in the W's generalized eignenspace $E(W) \in \mathbb{C}^{n \times n}$, since W is not symmetric.

For a connected undirected graph $\mathcal{G}$, if we see each undirected link of the graph $\mathcal{G}$ as two directed links with opposite directions and different weights, as shown in Figure 4.2, then $\hat{L} = \Delta^{-1} L$ is the Laplacian matrix of a strongly connected digraph $\hat{\mathcal{G}}$. We choose $\delta_i > 1, \forall i \in I$ to ensure $d_{\max}(\hat{\mathcal{G}}) \leq d_{\max}(\mathcal{G})$, so that α will satisfy the maximum node degree constraint of graph $\hat{\mathcal{G}}$. For a strongly connected graph, the Perron matrix W has the simple eigenvalue $\lambda_1 = 1$ as the spectral radius $\rho(W)$; all the other eigenvalues of W are in a unit circle $|\lambda_i| < 1, i = 2, \dots, n$. Therefore, from (4.53), we obtain

$$\lim_{k \to \infty} W^k = J_1. \tag{4.54}$$

Since $\lambda_1 = 1$ is the simple eigenvalue, according to Lemma 4.1, the associated projector J_1 is given by

$$J_1 = \frac{1 \delta^T}{\delta^T 1}, \tag{4.55}$$

where δ defined in Equation (4.48). It means

$$\lim_{k \to \infty} W^k = \frac{1 \delta^T}{\delta^T 1}, \tag{4.56}$$

which leads to Equations (4.49) and (4.50).

Remark 4.6   Theorem 4.1 is a direct application of the famous Perron–Frobenius theorem [35]. Slightly different versions of this theorem are presented in Refs. [19] and [36]. Setting $\delta_i \geq 1, \forall i \in I$ is a sufficient condition to ensure convergence of the consensus algorithm.

Remark 4.7   Setting weights δ_i in the consensus algorithm makes the information flow rate imbalance between any pair of SU nodes. For any pair of neighboring SUs (v_i, v_j), the ith SU has the stepsize (ϵ/δ_i), while the jth SU has the stepsize (ϵ/δ_j). This makes the network matrix, Laplacian matrix and Perron

Figure 4.2 A two-node network as an example for symmetric and asymmetric information flow. The left pair of SU nodes have 1 as the identical weight, which leads to symmetric information flow and average consensus. The right pair of SU nodes has different weights, which leads to asymmetric information flow and weighted average consensus.

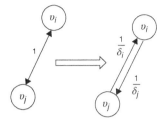

matrix, asymmetric, and the final convergence value deviates from the average consensus. Intuitively speaking, in Figure 4.2, the pair of SUs on the left has equal link weights and same stepsize α. In the iteration process, $\forall i, j$, the statistics held by the ith SU and jth SU converge to each other in the same rate, and the iteration stops when they meet at the average of both initial values, which leads to the average consensus convergence of the whole network. In contrast, the pair of SUs on the right has different link weights and different stepsizes. In the iteration process, statistics held by the ith SU and jth SU converge to each other in different rates, which makes the final consensus value deviate from the average consensus and converge to a linearly weighted average of the initial measurement statistics of the whole SU network.

Remark 4.8   Setting $\Delta = I$, all the weight $\delta_i = 1, \forall i$, we have W as a symmetric matrix with real eigenspectrum and eigenspace. 1 is the simple and largest eigenvalue of W; the vector $\mathbf{1}$ and $\mathbf{1}^T$ are the associated left and right eigenvectors, respectively. The convergence of the consensus iteration is given as

$$\lim_{k \to \infty} W^k x(0) = \frac{\mathbf{1}\mathbf{1}^T}{\mathbf{1}^T \mathbf{1}} x(0) = \frac{\sum_{i=1}^{n} x_i(0)}{n} \mathbf{1}, \tag{4.57}$$

which is the average consensus algorithm extensively studied in the literature [20, 24, 37], to name a few.

4.4.2 Dynamic Communication Channels

Realistic SU networks suffer from noise and error interruption or power use constraints. Link failures and dynamic switching communication channels should be considered. In this section, we characterize the conditions for the weighted average consensus convergence on the dynamic communication channels.

For a network of n SUs, there are a finite number, say a total of r, of possible communication graphs. We denote the set of all possible graphs by $\{G_1, \dots, G_r\}$, and the set of corresponding Laplacian matrices and Perron matrices given by $\{L_1, \dots, L_r\}$ and $\{W_1, \dots, W_r\}$, respectively. We have

$$W_s = I - \alpha \Delta^{-1} L_s, \tag{4.58}$$

for any $1 \leq s \leq r$. The weighted average consensus algorithm is given by

$$x(k+1) = W_{s(k)}x(k),$$ (4.59)

where the indices $s(k)$ are integers and satisfy $1 \leq s(k) \leq r$ for all $k > 0$. Here, we use the notion $W_{s(k)}$ to denote the graph sequence in the iteration because the graph sequences could be stochastic or deterministic. We will $W(k)$ to denote the stochastic case for the rest of chapter.

Theorem 4.2   For the iteration process (4.59), if the stepsize α satisfies

$$0 < \alpha < \frac{1}{n},$$ (4.60)

where n is the number of the SU nodes in the network, and the elements of matrix $\Delta = \mathrm{diag}\{\delta_1, \dots, \delta_n\}$ satisfy $\delta_i \geq 1, \forall i \in \mathcal{I}$, and the collection of bidirectional communication graphs that occur infinitely often are jointly connected, then the iteration converges to

$$\lim_{k \to \infty} x_i(k) = \frac{\sum_{i=1}^{n} \delta_i x_i(0)}{\sum_{i=1}^{n} \delta_i}, \forall i \in \mathcal{I}.$$ (4.61)

Proof: We show that consensus iteration (4.59) is actually a paracontraction process under the $\mathcal{L}_\infty$ norm and its fixed point is decided by the eigenspaces of the related Perron matrices.

A matrix $M \in \mathcal{R}^{n \times n}$ is called paracontracting [38] with respect to a vector norm $\| \cdot \|$ if

$$Mx \neq x \Leftrightarrow \|Mx\| < \|x\|.$$ (4.62)

For a matrix M, we denote $\mathcal{H}(M)$ as its fixed-point subspace, that is, $\mathcal{H}(M) = \{x | x \in \mathbb{R}^n | Mx = x\}$. Apparently, $\mathcal{H}(M)$ is M's eigenspace associated with the eigenvalue 1.

Before presenting the main proof, we give Lemmas 4.2–4.4 as follows:

Lemma 4.2 *([38])*   Suppose that a finite set of square matrices $\{W_1, \dots, W_r\}$ are paracontracting. Let $\{i(k)\}_{i=0}^{\infty}$, with $1 \leq i(k) \leq r$, be a sequence of integers, and denote by $\mathcal{J}$ the set of all integers that appear infinitely often in the sequence. Then for all $x(0) \in \mathbb{R}^n$ the sequence of vectors $x(k+1) = W_{i(k)}x(k), k \leq 0$, has a limit $x^* \in \bigcap_{i \in \mathcal{J}} \mathcal{H}(W_i)$.

Lemma 4.3 *([20])*   If a collection of graphs $\{G_1, \dots, G_p\}$ are jointly connected, then their corresponding Perron matrices satisfy

$$\bigcap_{i=1}^{p} \mathcal{H}(W_i) = \mathcal{H}\left(\frac{1}{p}\sum_{i=1}^{p} W_i\right) = \mathrm{span}(\mathbf{1}).$$ (4.63)

The proof of Lemma 4.3 follows the same procedure in the proof of Lemma 4.2 in [20]. For the jointly connected collection of possible graphs $\{G_1, \ldots, G_r\}, r \geq p$, we have

$$\bigcap_{i=1}^{r} \mathcal{H}(W_i) = \bigcap_{i=1}^{p} \mathcal{H}(W_i) = \text{span}\{\mathbf{1}\}. \tag{4.64}$$

Lemma 4.4 For any possible graph G, the associated graph Perron matrix is $W = I - \alpha\Delta^{-1}L$; we have $\|W\|_\infty \leq 1$. For any jointly connected graph sequence $\{G_1, \ldots, G_p\}$, then the matrix

$$\tilde{W} = \prod_{i=1}^{p} W_i \tag{4.65}$$

is a paracontracting matrix having $\mathbf{1}$ as the right eigenvector associated with the simple eigenvalue 1.

To prove Lemma 4.4, we first show that $\|\tilde{W}\|_\infty \leq 1$, which is equivalent to the fact that the maximum value in the network is nonincreasing and the minimum value in the network is nondecreasing. Under any possible undirected graph $\mathcal{G}$ and the associated Perron matrix $W = I - \epsilon\Delta^{-1}L$ defined in (4.45), if we assume the ith SU holds the maximum value in the network, we have the algorithm (4.21) as

$$x_{\max}(k+1) = x_{\max}(k) + \frac{\alpha}{\delta_i} \sum_{j \in N_i} (x_j(k) - x_{\max}(k))$$

$$= \left(1 - \alpha\frac{|N_i|}{\delta_i}\right) x_{\max}(k) + \frac{\alpha}{\delta_i} \sum_{j \in N_i} x_j(k)$$

because $0 < \alpha < 1/n, n \geq N_i$ and $\delta_i > 1$, we have $0 < (|N_i|\alpha/\delta_{\max}) < 1$, which means $x_{\max}$ is non-increasing in every step of the iteration and $x_{\max}$ always stays in the convex hull formed by $x_{\max}$ and its local neighbors, no matter how the graphs are sequenced. Following the same procedure, we can prove $x_{\min}$ is non-decreasing in every step of the iteration and $x_{\min}$ always stays in the convex hull formed by $x_{\min}$ and its local neighbors. Therefore, we have $\|W\|_\infty \leq 1$ which leads to $\|\tilde{W}\|_\infty \leq \prod_{i=1}^{q} \|W_i\|_\infty \leq 1$.

Second, we prove that $\tilde{W}$ is paracontracting. According to Lemma 4.3, we know $\tilde{W}$ has a simple eigenvalue 1 associated with the eigenvector $\mathbf{1}$. Therefore, in the iteration $x(k+p) = \prod_{i=1}^{p} W_i x(k) = \tilde{W}x(k)$, for $x(k) \notin \text{span}(\mathbf{1})$, then $\max\{|x_{\max}|, |x_{\min}|\}$ is strictly decreasing, which means $\|W_{q_j} x\|_\infty < \|x\|_\infty$. If, say, $|x_{\max}| > |x_{\min}|$ and $x_{\max}$ remains the same after the iteration, then the vector $[0, \ldots 0, x_{\max}, 0, \ldots, 0]^T$ is a also an eigenvector associated with the eigenvalue 1, which means 1 is not a simple eigenvalue of $\tilde{W}$ and it contradicts with Lemma 4.3. Thus, we have $\tilde{W}$ is paracontracting according to the definition Equation (4.62). This finishes the proof of Lemma 4.4.

Under the condition that the collection of the jointly connected graphs occurs infinitely, we can write the matrix sequence as $\prod_{i=1}^{k} W_{s(k)} = \prod_{j=1}^{h} \tilde{W}_j$, the associated graph sequences $\{G_1, \ldots, G_{p_j}\}$, $\sum_{j=1}^{h} p_j = k$ are all jointly connected and $\tilde{W}_j = \prod_{l=1}^{p_j} W_l$ are paracontracting for all $1 \leq j \leq h$ according to Lemma 4.4. So $\prod_{i=1}^{k} W_{s(k)}$ is an infinite paracontracting process as $k \to \infty$.

To prove the final convergence value, we decompose the generalized eigenspace $\mathbb{E}^n$ of $\tilde{W}$ as $E_1 \bigoplus E_s$ [39], where $\tilde{W}$ is defined in Equation (4.65) and $E_1 = \text{span}\{\mathbf{1}\}$ associated with the simple eigenvalue 1, and E_s is the collection of all the other eigenspaces. For any initial value $x(0)$, we decompose $x(0) = u(0) + w(0)$, where $u(0) \in E_1$ and $w(0) \in E_s$. Because $u(0) \in \bigcap_{i=1}^{r} \mathcal{H}(W_i)$, the sequence given by $u(k+1) = W_{s(k)}u(k)$ is constant, and the limit $u^* = u(0)$. On the other hand, because E_s is invariant under all $W_{s(k)}$ [39], the sequence of vectors given by $w(k+1) = W_{s(k)}w(k)$ all belong to E_s, which leads to the limit $w^* \in E_s$. From the properties of the infinite paracontracting process (see Lemma 4.2), w^* has to be in the fixed subspace of $\tilde{W}$. It means $w^* \in \text{span}\{\mathbf{1}\} = E_1$, which yields $w^* = 0$, because $E_1 \bigcap E_s = \emptyset$. Therefore, the iterative process (4.59) has the limit $x^* = u^*$, and u^* is given by the spectral projection of matrix $\tilde{W}$, associated with the simple eigenvalue 1, according to Lemma 4.1, $x^* = u^* = \frac{\mathbf{1}\delta^T}{\delta^T\mathbf{1}}x(0) = \frac{\sum_{i=1}^{n} \delta_i x_i(0)}{\sum_{i=1}^{n} \delta_i}\mathbf{1}$, where $\delta = [\delta_1, \ldots, \delta_n]^T$ and δ_i is the element of the diagonal matrix Δ. This finishes the proof of Theorem 4.2.

Remark 4.9 Following the spectral decomposition, the jointly connected condition ensures the infinite matrix product

$$\lim_{k \to \infty} W_{s(k)}^k = \lim_{k \to \infty} \tilde{W}^k = \frac{\mathbf{1}\delta^T}{\delta^T\mathbf{1}}, \tag{4.66}$$

where $\tilde{W}$ is defined Equation (4.65) and $\tilde{W}$ has a simple eigenvector $\mathbf{1}$, associated with the eigenvalue $\lambda_1 = 1$, and all the other eigenvalues stays in the unit circle, $|\lambda_i| < 1, \forall i = 2, \ldots, n$.

Remark 4.10 Theorem 4.2 encompasses the average consensus as a special case when $\delta = I$ and $W_{s(k)}$ are symmetric matrices. For symmetric $W_{s(k)}$, we have $\|W_{s(k)}\|_2 = \rho(W_{s(k)}) \leq 1$, based on which the convergence analysis is given in [20]. For asymmetric $W_{s(k)}$, we adopt the $\mathcal{L}_\infty$ norm $\|W_{s(k)}\|_\infty$ for the convergence analysis. Meanwhile, the fixed communication topology, Theorem 4.1, is a special case when $W_{s(k)} = W, \forall k \geq 0$.

Remark 4.11 Theorem 4.2 requires weak long-term connectivity which contains both deterministic and stochastic time-varying graph sequences, and the convergence rate in general may not exist. If we further assume each link has an independent probability to fail, for example, the link erasure model [23], we can give in the next subsection an estimation of the convergence rate of the consensus iteration.

4.4.3 Convergence Rate with Random Link Failures

For a SU network denoted as $\mathcal{G} = (\mathcal{E}, \mathcal{V})$, we assume $\mathcal{G}$ is a connected undirected graph and $\mathcal{E}$ is the set of realizable edges. We assign each pair of neighboring SUs the online and offline probabilities at each time step as P_{ij} and $1 - P_{ij}$, respectively. Then, at the arbitrary time index k, the network of n SUs is modeled by the graph $\mathcal{G}(k) = (E(k), \mathcal{V})$, where $E(k)$ denotes the edge set at time k.

Then the consensus iteration (4.59) becomes a random process, and it is modeled as

$$x(k + 1) = W(k)x(k), \tag{4.67}$$

where $W(k)$ is defined as

$$W(k) = I - \alpha \Delta^{-1} L(k), \tag{4.68}$$

where $\Delta = \mathrm{diag}\{\delta_1, \ldots, \delta_n\}$ satisfies $\delta_i \geq 1, \forall i \in \mathcal{I}$, $W(k)$ and $L(k)$ are the Perron matrix and Laplacian matrix, respectively, of the dynamic communication graph $\mathcal{G}(k)$ at time k. We assume that the link failures among the SU network happen independently, so all $L(k)$'s and $W(k)$'s are independent and identically distributed. We have Lemma 4.5 as follows:

Lemma 4.5 If the SU network forms a connected undirected communication graph $\mathcal{G} = (\mathcal{E}, \mathcal{V})$, each link $e_{ij} \in \mathcal{E}$ has the online and offline probability as P_{ij} and $1 - P_{ij}$, where $P_{ij} \in (0, 1)$, the stepsize α in Equation (4.68) satisfies the maximum node degree constraint $0 < \alpha < \frac{1}{d_{\max}(\mathcal{G})}$, then the vector sequence $\{x(i)\}_{i=0}^{\infty}$ in (4.21) converges exponentially in the sense that

$$\lim_{k \to \infty} \| \mathrm{E}\,(x(k)) - x^*\mathbf{1}\|_2 = 0. \quad \forall x(0) \in \mathbb{R}^{n \times 1} \tag{4.69}$$

The decay factor of the convergence is given by $\rho(\overline{W} - J_1)$, where $0 < \rho(\overline{W} - J_1) < 1$ is the spectral radius of $\overline{W} - J_1$, $\overline{W} = E(W)$, and J_1 is defined in Equation (4.55).

Proof: We have the error dynamics of the algorithm (4.67) as

$$x(k + 1) - x^*\mathbf{1} = W(k)x(k) - J_1 x(0),$$

$$= \prod_{j=0}^{k} W(j)x(0) - J_1 x(0). \tag{4.70}$$

Since δ and $\mathbf{1}$ are, respectively, the left and right eigenvetor of $W(k), \forall k \geq 0$, associated with the eigenvalue $\lambda_1 = 1$, we have $W(k)J_1 = J_1$ and $J_1 W(k) = J_1, \forall k \geq 0$, which yield

$$x(k + 1) - x^*\mathbf{1} = W(k) \prod_{j=0}^{k-1} W(j)x(0) - J_1 \prod_{j=0}^{k-1} W(j)x(0),$$

$$= W(k)x(k) - J_1 x(k)$$

$$= \left(W(k) - J_1 \right) x(k). \tag{4.71}$$

Since $W(k)J_1 = J_1$ and $J_1 J_1 = J_1$, we have

$$x(k+1) - x^* \mathbf{1} = \left(W(k) - J_1 \right) \left(x(k) - J_1 x(0) \right)$$

$$= \prod_{j=0}^{k} \left(W(k) - J_1 \right) \left(x(0) - J_1 x(0) \right). \qquad (4.72)$$

Since all $L(k)$'s and $W(k)$'s are independent and identically distributed, we have

$$\mathrm{E}\,(x(k+1) - x^* \mathbf{1}) = \mathrm{E}\left(\prod_{j=0}^{k} \left(W(k) - J_1 \right) \right) \mathrm{E}\,(x(0) - x^* \mathbf{1})$$

$$= \prod_{j=0}^{k} \mathrm{E}\left(W(k) - J_1 \right) (\mathrm{E}(x(0)) - x^* \mathbf{1})$$

$$= \left(\overline{W} - J_1 \right)^k (\mathrm{E}(x(0)) - x^* \mathbf{1}), \qquad (4.73)$$

which yields

$$\| \mathrm{E}(x(k+1)) - x^* \mathbf{1} \|_2 = \| \left(\overline{W} - J_1 \right)^k \|_2 \| (\mathrm{E}(x(0)) - x^* \mathbf{1}) \|_2.$$

For $\overline{W} = \mathrm{E}(W)$, we have

$$\overline{W} = I - \epsilon \Delta^{-1} L_p, \qquad (4.74)$$

where L_p is from the link probability, defined as follows

$$l_{p_{ij}} = \begin{cases} \sum_{j=1}^{n} P_{ij}, & \text{if } i = j, \\ -P_{ij} & \text{if } i \neq j, \text{ and } (v_i, v_j) \in \mathcal{E} . \\ 0 & \text{otherwise} \end{cases} \qquad (4.75)$$

We can see that L_p is still a Laplacian matrix of a connected graph with P_{ij} as its link weights. For a strongly connected graph, we have $\rho(\overline{W} - J_1) < 1$, when α satisfies the maximum node degree constraint.

According to the famous Gelfand's formula, $\| (\overline{W} - J_1)^k \|_2$ has the same growth rate as $\rho(\overline{W} - J_1)^k$ as $k \to \infty$, which leads to the exponential convergence of (4.69), and the decay factor is $\rho(\overline{W} - J_1)$. □

Remark 4.12 The decay factor for the convergence rate is the network connectivity $\rho(\overline{W} - J_1)$ which relates to the network topology and the link weights, as well as the link failure probability matrix L_p. For optimizing the convergence rate, interested readers can refer to Refs. [23, 24, 40].

Practically, it's unnecessary for the SU network to reach the limit in the consensus iteration. We can derive the upper bound on the iteration number at which all SUs are ϵ close to the final convergence value in the probability sense, which is called ϵ-convergence in Ref. [26].

Theorem 4.3   Under the same condition of Lemma 4.5, $\forall \epsilon > 0$ and $k \geq T(\epsilon)$, for the iteration (4.21), we have

$$\Pr\{\max_{1\leq i\leq n}|x_i(k) - x^*| \geq \epsilon | \mathcal{H}_k\} \leq \epsilon, \quad k \in \{0, 1\} \tag{4.76}$$

and

$$T(\epsilon) \leq \frac{3/2 log \epsilon^{-1} + 1/2 log(K)}{1 - E\left(\|(W - J_1)\|_\infty\right)}, \tag{4.77}$$

where

$$K = \sum_{i=1}^{n}\left(2m\sigma_i^4 + 4E_s|h_i|^2\sigma_i^2 + (m\sigma_i^2 + E_s|h_i|^2)^2\right) \tag{4.78}$$

and J_1 is defined in (4.55), σ_i is the measurement noise variance for the ith SU, and E_s is the signal energy defined in (4.6) and h_i is the channel gain.

Proof: Since $\max_{1\leq i\leq n}|x_i(k) - x^*| = \|x(k) - x^*\|_\infty$, we have

$$\Pr\{\|x(k) - x^*\mathbf{1}\|_\infty \geq \epsilon\, \mathcal{H}_k\}$$

$$= \Pr\{\|x(k) - x^*\mathbf{1}\|_\infty^2 \geq \epsilon | \mathcal{H}_k^2\} \leq \frac{E\{\|x(k) - x^*\mathbf{1}\|_\infty^2 | \mathcal{H}_k^2\}}{\epsilon^2}, \tag{4.79}$$

where the second equation is from the Markov inequality. According to the paracontraction norm properties of $\mathcal{H}_\infty$-norm, from Eqn. (4.71), we have

$$\|x(k) - x^*\mathbf{1}\|_\infty^2 \leq \prod_{i=1}^{k}\|(W(k) - J_1)\|_\infty^{2k}\|x(0)\|_\infty^2. \tag{4.80}$$

Since $W(k)$'s are identically and independently distributed, we have

$$E(\|x(k) - x^*\mathbf{1}\|_\infty^2) \leq E\left(\|W - J_1\|_\infty\right)^{2k} E(\|x(0)\|_\infty^2). \tag{4.81}$$

If we choose a vector $\tilde{x}$ that $\|\tilde{x}\|_\infty = 1$ and $\delta^T\tilde{x} = 0$, where δ is defined in Equation (4.48), we have $J_1\tilde{x} = 0$ and following Lemma 4.4 in the proof of Theorem 4.2, we have

$$\|(W(k) - J_1)\tilde{x}\|_\infty = \|W(k)\tilde{x}\|_\infty \leq \|\tilde{x}\|_\infty \tag{4.82}$$

when $W(k)$ has 1 as a simple eigenvalue, we have

$$\|(W(k) - J_1)\tilde{x}\|_\infty < \|\tilde{x}\|_\infty, \tag{4.83}$$

which means

$$\|(W(k) - J_1)\|_\infty = \max_{\|\tilde{x}\|_\infty=1}\frac{\|(W(k) - J_1)\tilde{x}\|_\infty}{\|\tilde{x}\|_\infty} \leq 1 \tag{4.84}$$

and

$$E\left(\|(W - J_1)\|_\infty\right) < 1, \tag{4.85}$$

we drop the index of W because $W(k)$ are identically distributed. We also have

$$\|x(0)\|_\infty^2 \leq \|x(0)\|_2^2. \tag{4.86}$$

Substituting (4.80) and (4.86) into (4.79), we have

$$\Pr\{\|x(k) - x^*\mathbf{1}\|_\infty \geq \epsilon | \mathcal{H}_k\} \leq \frac{\mathrm{E}\left(\|(W - J_1)\|_\infty\right)^{2k} \mathrm{E}\{\|x(0)\|_2\}}{\epsilon^2}. \tag{4.87}$$

Let

$$\frac{\mathrm{E}\left(\|(W - J_1)\|_\infty\right)^{2k} \mathrm{E}\{\|x(0)\|_2\}}{\epsilon^2} = \epsilon. \tag{4.88}$$

We obtain

$$T(\epsilon) = \frac{3/2 \log \epsilon^{-1} + 1/2 \log(\mathrm{E}\{\|x(0)\|_2^2 | \mathcal{H}_k\})}{- \log(\mathrm{E}\left(\|(W - J_1)\|_\infty\right)} \tag{4.89}$$

Therefore, we have

$$T(\epsilon) = \frac{3/2 \log \epsilon^{-1} + 1/2 \log(\mathrm{E}\{\|x(0)\|_2^2 | \mathcal{H}_k\})}{- \log(\mathrm{E}\left(\|(W - J_1)\|_\infty\right)}. \tag{4.90}$$

From the inequality $\log(1 + u) \leq u$ when u is small, let $1 + u = \mathrm{E}\left(\|(W - J_1)\|_\infty\right)$, we obtain

$$- \log \mathrm{E}\left(\|(W - J_1)\|_\infty\right) \geq -u = 1 - \mathrm{E}\left(\|(W - J_1)\|_\infty\right).$$

Thus, we have

$$T(\epsilon) \leq \frac{3/2 \log \epsilon^{-1} + 1/2 \log(\mathrm{E}\{\|x(0)\|_2^2 | \mathcal{H}_k\})}{1 - \mathrm{E}\left(\|(W - J_1)\|_\infty\right)}. \tag{4.91}$$

Meanwhile, according to (4.8) and (4.9), we have

$$\begin{aligned}
\mathrm{E}(\|x(0)\|_2^2 | \mathcal{H}_0) &= \sum_{i=1}^{n} \mathrm{E}(x_i^2(0) | \mathcal{H}_0) \\
&< \sum_{i=1}^{n} \mathrm{E}(x_i^2(0) | \mathcal{H}_1) \\
&= \sum_{i=1}^{n} \left(\mathrm{Var}(x_i(0) | \mathcal{H}_1) + \mathrm{E}^2(x_i(0) | \mathcal{H}_1)\right) \\
&\leq \sum_{i=1}^{n} \left(2m\sigma_i^4 + 4E_s|h_i|^2\sigma_i^2 + (m\sigma_i^2 + E_s|h_i|^2)^2\right).
\end{aligned} \tag{4.92}$$

where σ_i is the measurement noise variance for the ith SU, and E_s is the signal energy defined in (4.6) and h_i is the channel gain. We obtain 4.77. □

Remark 4.13   The ϵ-convergence of the average consensus or gossip algorithm has been extensively studied in Refs. [16, 26, 41]. Theorem 4.3 is a generalization to weighted average consensus convergence with random link failures. From (4.77), we can see clearly that the convergence rate of ϵ-convergence depends on the desired accuracy ϵ, the measurement channel noise variance σ_i, signal energy E_s, channel gain h_i, and the expectation $\mathrm{E}\left(\|(W - J_1)\|_\infty\right)$.

Remark 4.14   Practically, $\mathrm{E}\left(\|(W - J_1)\|_\infty\right)$ is not easy to compute. Because the norm $\|\cdot\|$ is a convex function, we have

$$\mathrm{E}\left(\|(W - J_1)\|_\infty\right) \geq \|(\overline{W} - J_1)\|_\infty$$
$$\geq \rho(\overline{W} - J_1). \tag{4.93}$$

The second inequality is from the property of the matrix spectral radius. Therefore, we can use $\rho(\overline{W} - J_1)$ as an estimation of the minima of $\mathrm{E}\left(\|(W - J_1)\|_\infty\right)$ so that we have an approximation of $T(\epsilon)$.

4.5   Simulations and Performance Evaluation

In this section, we conduct simulations to study the performance of our proposed distributed weighted combining scheme. The simulation setup, evaluation metrics, and comparison methodologies are described. We show the convergence of the weighted consensus algorithm, and Monte Carlo simulation [32] is also conducted to evaluate the detection performance of the proposed distributed spectrum sensing scheme.

4.5.1   SU Network Setup

In the simulation, we mainly consider a 10-node SU network shown in Figure 4.3, with both fixed and dynamic communication channels. In particular, Figure 4.3(b) shows that 10 SUs communicate with each other through fixed duplex communication channels, and Figure 4.3(b) considers a dynamic communication channel case which is generated from Figure 4.3(b) by assigning a failure probability to each communication link. We use Fig. 4.3(b) as a general model for the realistic SU networks suffering from noise and error interruptions. In the later simulations with network size as a variable, we also consider 20-node and 30-node SU networks, which are modeled by 4×5 and 5×6 grid graphs, respectively.

In the sensing stage, all SUs are assumed to be static and have uncorrelated measuring channels with independent fading effects. We mainly consider AWGN channel and Rayleigh fading channel in the simulation. According to the energy detector model, each SU generates the measurement Y_i of the PU's signal energy under the hypothesis $\mathcal{H}_1$ in (4.4), with the time-bandwidth product $m = 12$. In the fusion stage, each SU sets its initial measurement

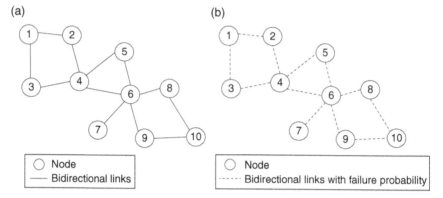

Figure 4.3 Communication network of 10 SUs. (a) SU network with fixed communication duplex channels and (b) snapshot of the SU network with dynamic communication channels with random link failure probability.

statistic $x_i(0) = Y_i$ and starts the measurement fusion by the algorithm in Equation (4.21) until the final consensus is reached after convergence. The measurement fusion only adopts local communications without fusion center. In both sensing and fusion stage, all SUs are running the synchronized clock during the sensing process and the consensus iteration, which is not a strong assumption also used in Ref. [14].

4.5.2 Convergence of Weighted Average Consensus

Consensus convergence is crucial in the distributed measurement fusion process. Figure 4.4 shows the convergence performance of the proposed algorithm with respect to the iteration step under both fixed and dynamic communication channels. The optimal weight (4.33) is set on each SU. We found in Figure 4.4(a) that within 30 steps, the differences of x_i among all the SUs with fixed communication channels are less than 1 dB, indicating the consensus has been reached on the global decision statistics 11.1 dB. Figure 4.4(b) shows the convergence of the algorithm under dynamic communication channels with independent link failures, where the failure probability of each link to be 0.4. We observed that the consensus is also achieved within 30 iteration steps and the random communication link failure does not affect the convergence of the fusion algorithm.

In order to characterize the convergence performance in terms of the detection probability P_d, Figure 4.5 shows the trend of P_d curve during the fusion process with respect to the consensus iteration time step under fixed communication channels. We can find that the detection probability of 10 SU nodes converges to the same value from centralized (WGC) approach within 35 steps. The false alarm is set at $P_f = 0.2$, the variance of Gaussian noise $\sigma_i = 1, \forall i$, and the channel SNR varies from 0 to -15 dB.

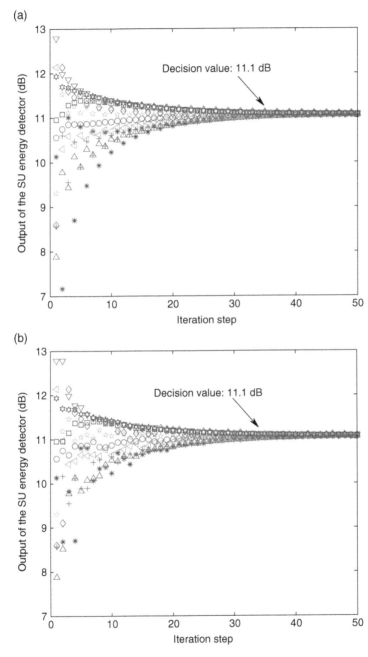

Figure 4.4 Convergence of the proposed consensus algorithm under the 10-node SU network in Figure 4.3. The output of the energy detector is *m* times the actual received PU energy, and *m* is the time-bandwidth product. (a) Fixed communication channels, $\epsilon = 0.19$. (b) Dynamic communication channels with independent link failure probability, $\epsilon = 0.19$. Link failure probability = 0.4 and convergence probability = 1.

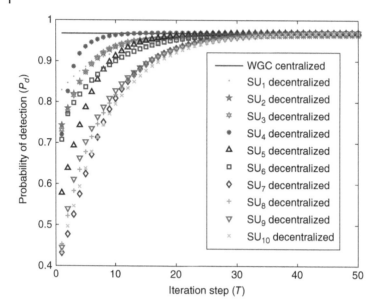

Figure 4.5 Detecting probability P_d with respect to iteration time step, $\alpha = 0.19$, fixed communication topology, SU number $N = 10$, $P_f = 0.2$, channel SNR at the output of the energy detection ranges from 0 dB to -15 dB.

4.5.3 Metrics and Methodologies

To compare the performance of different detection approaches, such as distributed weighted combing (DWGC) with the centralized weighted combining, the decentralized equal gain combining (EGC) [14], and hard bit combining OR-rule scheme (OR) [7], the metrics P_d (detection probability) and P_f (probability of false alarm) are taken in consideration. A high P_d will result in high P_f, which increases the interference to PUs. On the other hand, a low P_f will result in low P_d and lead to low spectrum utilization.

We vary the measurement channel SNR to show the detecting performance under different channel conditions, AWGN and Rayleigh fading channel. It is challenging to achieve high detecting performance under the condition of low SNR. The threshold λ is computed from the false alarm constraints under the hypothesis $\mathcal{H}_0$ in Equation (4.4). Soft combining schemes (DWGC and EGC) share the same threshold, since the measurement output under $\mathcal{H}_0$ is independent of the average SNR when the PU signal is absent. The decision threshold of OR rule is decided by the largest measurement among the SU network under the hypothesis $\mathcal{H}_0$.

In Section 4.5.4, we compare detection performance of the proposed DWGC with the centralized weighted combining, the decentralized EGC [14], and hard bit combining OR [7], with respect to the measurement channel conditions, PU transmission power and network sizes. During the performance evaluation, we mainly consider the AWGN measurement channel which is an approximation of general fading channels. Meanwhile, we also evaluate the effect of the Rayleigh fading on the detecting performance, and we compare the proposed optimal weighted design DWGC with the special heuristic weights (AWGC) in Ref. [31] based on estimated average channel SNR.

4.5.4 Performance Evaluation

4.5.4.1 Receiver Operating Curves under AWGN Channels

Figure 4.6 shows the receiver operating curves (ROCs) under AWGN channel for the 10-node SU network shown in Figure 4.3, where channel SNR of the SU network ranges from 0 to −15 dB. The proposed DWGC achieves the best performance which is comparable with the centralized WGC with both fixed and dynamic communication channel. As expected, the temporal communication channel failures do not affect the detection performance of DWGC approach. The EGC approach has satisfactory performance but still worse than the WGC approaches. OR rule performs the worst. Particularly, At

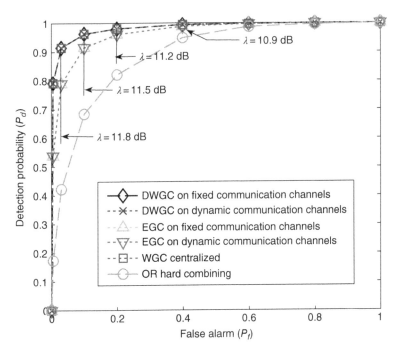

Figure 4.6 ROC of the 10-node SU network under AWGN channel, channel SNR ranges from −0 to −10 dB, and λ is the decision threshold.

the false-alarm constraint $P_f = 0.2$, the DWGCs offer the detection probability $P_d = 0.97$, which is in accordance with the numerical results shown in Figure 4.5. EGC and OR give $P_d = 0.95$ and $P_d = 0.82$, respectively. Figure 4.6 shows clearly the proposed DWGC scheme achieves comparable performance with centralized weighted combining design and outweighs EGC and OR-rule combining.

4.5.4.2 ROCs under Rayleigh Fading Channels

Figure 4.7 shows the ROC curves under Rayleigh fading channel for DWGC, EGC, centralized WGC, and OR-rule and AWGC approaches. In comparison with the performance under AWGN channel in average sense, the average SNR of the Rayleigh fading is also set at −5 dB. We can observe weighted combing schemes are still robust under Rayleigh fading, but EGC approach and OR-rule approach suffer from the fading and the performance degrade severely. DWGC has the best performance under both fixed and dynamic communication channels. The AWGC-based heuristic weights design in Ref. [31] offers comparable performance with the optimal weighted design. The DWGC and AWGC achieve comparable performance with centralized

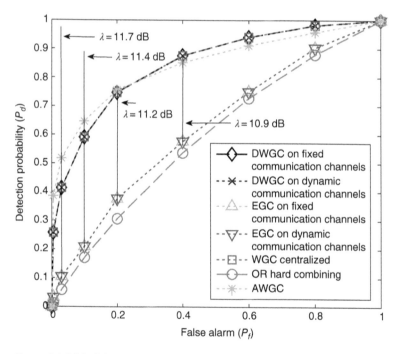

Figure 4.7 ROC of the 10-node SU network under Rayleigh fading channel. All SUs have same average channel SNR −5 dB.

WGC. At the false-alarm constraint $P_f = 0.2$, DWGC achieves detection probability $P_d = 0.76$ and the performance is comparable with AWGC and centralized WGC, while EGC only offers $P_d = 0.38$ and OR rule is the worst with $P_d = 0.31$. Comparing Figures 4.6 and 4.7, we see that the performance degradation of the DWGC and centralized WGC due to Rayleigh fading channel effects is within 20%, while the performance of EGC and OR rule degrades more than 50% after the Rayleigh fading is considered.

4.5.4.3 ROCs with Respect to the PU Transmission Power

In Figure 4.8, we compare the detection performance P_d of the proposed DWGC with existing schemes when PU transmission power varies. Under AWGN channel, since the variance of Gaussian noise is fixed at $\sigma_i = 1, \forall i$, the SNR η could reflect the signal received signal strength from PU. The DWGC always achieves the highest detection probability under both fixed and dynamic communication channels, and has comparable performance with centralized WGC. Especially when the PU transmission power is low, DWGC approach offers higher detection probability than EGC and OR-rule

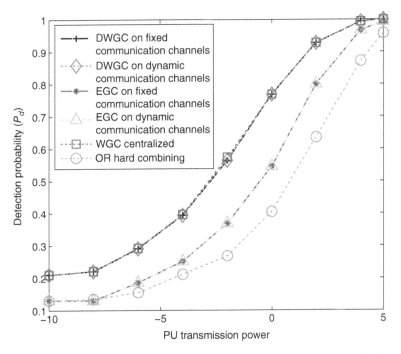

Figure 4.8 Detecting probability P_d with respect to PU signal SNR, $\epsilon = 0.19$, fixed communication topology, SU number $N = 10$, and $P_f = 0.02$; SUs are located 50–150 m away from PU.

approaches. Particularly, when channel SNR is 0 dB, the WGC approaches achieve detection probability of 0.77, which has 40 and 88% improvement than EGC approach and OR rule, respectively. When the PU channel SNR becomes larger, such as close to 5 dB, the three approaches offer similar performance.

4.5.4.4 ROCs with Respect to SU Network Sizes
In Figure 4.9, the ROC curves of the DWGC, EGC, OR rule, and centralized WGC, under AWGN measuring channel with different SU network sizes are presented. As shown in Figure 4.9(a) and (b), the SU network with 20 and 30 nodes are employed for performance evaluation under AWGN channel with identical channel condition. DWGC achieves the best performance under different network sizes, and have comparable performance with centralized WGC approach. In particular, within the false-alarm constraint $P_f \in [0.01, 0.1]$, DWGC achieves detection probability above 0.9. When the network size increases, the detection probability also increases. For both 20-node and 30-node cases, detection probability of DWGC is 10% higher than EGC and 25% higher than OR-rule approaches. Here, the variance of Gaussian noise is fixed at $\sigma_i = 1, \forall i$, and the measuring channel SNR η of the SU network ranges from 0 to −15 dB.

In Figure 4.10, we plot the ROC curves for DWGC, AWGC, EGC, OR rule, and centralized WGC under Rayleigh fading channel with network size with 20 and 30 nodes, respectively. We can see the proposed DWGC achieves comparable performance with centralized WGC, and the performance is much less affected by the Rayleigh fading than EGC and OR rule. Moreover, DWGC improves the performance as the network size increasing, and EGC rule is much less sensitive to the network size compared to the AWGN measurement channel case in Figure 4.9. Particularly, when the network size increases from 20 nodes to 30 nodes with the false alarm at 0.2, the detection probability of DWGC and AWGC increases 10%, while the performance improvement of EGC and OR rule is within 5%. From Figure 4.9(b), AWGC gives better performance than DWGC, since AWGC is a special weights design based on Rayleigh fading channel average SNR estimation and DWGC weight design does not consider Rayleigh fading effects.

From the earlier discussion, we see that the proposed consensus-based weighted combining achieves comparable performance with the centralized weighted combining scheme and outweighs significantly the existing equal gain combining and OR-rule combining approaches. The proposed distributed optimal weight design based on AWGN approximation models shows its effectiveness with respect to channel SNR variation, low PU signal strength condition, and network size increase. Under Rayleigh fading channel, the proposed AWGN channel optimal weighted design achieves comparable performance compared to the special heuristic weighted design based on average channel SNR estimation.

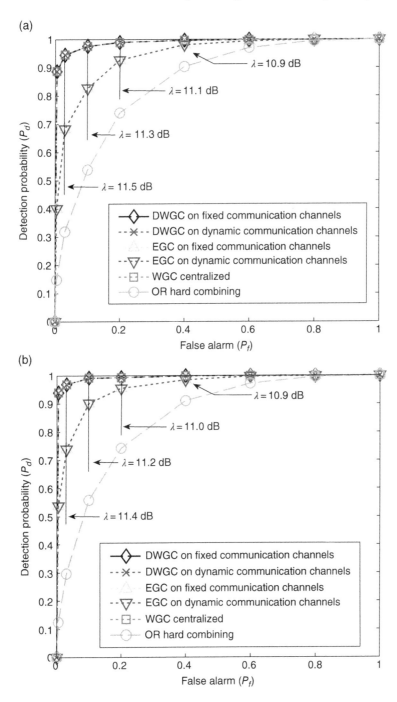

Figure 4.9 ROC under AWGN channel with different network sizes. (a) ROC of a 20-node SU grid network. The channel SNR ranges from 0 to −15 dB. (b) ROC of a 30-node SU grid network. The channel SNR ranges from 0 to −15 dB.

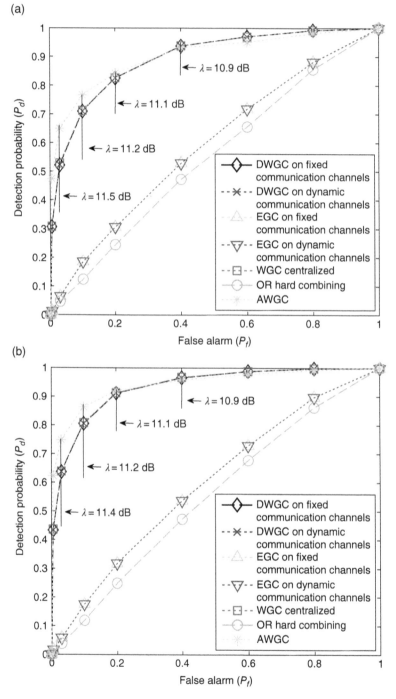

Figure 4.10 ROC under Rayleigh fading Channel with different network sizes. (a) ROC of a 20-node SU grid network. Channel average SNR ranges from −5 to −10 dB. (b) ROC of a 30-node SU grid network. Channel average SNR ranges from −5 to −10 dB.

4.6   Conclusion

In this chapter, we present a distributed weighted combining scheme for the cooperative spectrum sensing in CR networks. The proposed method is based on the weighted average consensus algorithm for both fixed and time-varying graphs. Through weighted local fusion iteration, each SU holds the global decision statistic from the weighted soft measurement combining throughout the network. We prove the convergence of the iteration and characterize the convergence rate of the consensus algorithm under independent link failure condition. We discuss the optimal weight design for the distributed soft combining and compare the results with the centralized optimal weighted combining. Simulation results show the proposed decentralized combining approach offers a comparable performance with the centralized weighted combining and outperforms existing average consensus-based equal gain combining schemes and the OR-rule hard bit combining scheme.

Notes

The results in this chapter were first published in Ref. [42].

Acknowledgment is given to ©2015 IEEE. Reprinted, with permission, from Zhang, W., Wang, Z., Liu, H., Guo, Y., Chen, Y., and Mitola, J. (2015) Distributed consensus-based weight design for cooperative spectrum sensing. IEEE Transactions on Parallel and Distributed Systems, 26 (1), 54–64.

References

1 Mitola, J. (2000) *An integrated agent architecture for software defined radio*, Ph.D. thesis, The Royal Institute of Technology, KTH.
2 Akyildiz, I.F., Lee, W.Y., Vuran, M.C., and Mohanty, S. (2006) NeXt generation/dynamic spectrum access/cognitive radio wireless networks: A survey. *Computer Networks*, **50**, 2127–2159.
3 Ma, J., Li, G., and Juang, B.H. (2009) Signal processing in cognitive radio. *Proceedings of the IEEE*, **97** (5), 805–823, doi:10.1109/JPROC.2009.2015707.
4 Yucek, T. and Arslan, H. (2009) A survey of spectrum sensing algorithms for cognitive radio applications. *IEEE Communications Surveys Tutorials*, **11** (1), 116–130, doi:10.1109/SURV.2009.090109.
5 Akyildiz, I.F., Lo, B.F., and Balakrishnan, R. (2011) Cooperative spectrum sensing in cognitive radio networks: A survey. *Physical Communication*, **4** (1), 40–62.
6 Cabric, D., Mishra, S., and Brodersen, R. (2004) Implementation issues in spectrum sensing for cognitive radios. *Conference Record of the Thirty-Eighth Asilomar Conference on Signals, Systems and Computers*, **1**, 772–776, doi:10.1109/ACSSC.2004.1399240.

7 Ghasemi, A. and Sousa, E. (2005) Collaborative spectrum sensing for opportunistic access in fading environments, in *First IEEE International Symposium on New Frontiers in Dynamic Spectrum Access Networks, DyS-PAN 2005*, pp. 131–136, doi:10.1109/DYSPAN.2005.1542627.

8 Ma, J. and Li, Y. (2007) Soft combination and detection for cooperative spectrum sensing in cognitive radio networks, in *IEEE Global Telecommunications Conference*, pp. 3139–3143, doi:10.1109/GLOCOM.2007.594.

9 Ben Letaief, K. and Zhang, W. (2009) Cooperative communications for cognitive radio networks. *Proceedings of the IEEE*, **97** (5), 878–893, doi:10.1109/JPROC.2009.2015716.

10 Zhang, W. and Letaief, K. (2008) Cooperative spectrum sensing with transmit and relay diversity in cognitive radio networks - [transaction letters]. *IEEE Transactions on Wireless Communications*, **7** (12), 4761–4766, doi:10.1109/T-WC.2008.060857.

11 Mishra, S., Sahai, A., and Brodersen, R. (2006) Cooperative sensing among cognitive radios, in *IEEE International Conference on Communications*, vol. 4, pp. 1658–1663, doi:10.1109/ICC.2006.254957.

12 Zou, Y., Yao, Y.D., and Zheng, B. (2011) A cooperative sensing based cognitive relay transmission scheme without a dedicated sensing relay channel in cognitive radio networks. *IEEE Transactions on Signal Processing*, **59** (2), 854–858.

13 Yu, F., Huang, M., and Tang, H. (2010) Biologically inspired consensus-based spectrum sensing in mobile ad hoc networks with cognitive radios. *IEEE Network*, **24** (3), 26–30, doi:10.1109/MNET.2010.5464224.

14 Li, Z., Yu, F.R., and Huang, M. (2010) A distributed consensus-based cooperative spectrum-sensing scheme in cognitive radio. *IEEE Transaction on Vehicular Technology*, **59** (1), 383–393.

15 Han, Z., Li, H., Yang, D., and Pei, C. (2011) Belief propagation based cooperative compressed spectrum sensing in wideband cognitive radio networks. *IEEE Transactions on Wireless Communications*, **10** (9), 3020–3031.

16 Yildiz, M.E., Aysal, T.C., and Barner, K.E. (2009) In-network cooperative spectrum sensing, in *EURASIP European Signal Processing Conference*, Glasgow, Scotland.

17 Tsitsiklis, J. (1984) *Problems in decentralized decision making and computation*, Ph.D. thesis, Department of Electrical Engineering and Computer Science, M.I.T., Boston, MA.

18 Jadbabaie, A., Lin, J., and Morse, A. (2003) Coordination of groups of mobile autonomous agents using nearest neighbor rules. *IEEE Transactions on Automatic Control*, **48** (6), 988–1001, doi:10.1109/TAC.2003.812781.

19 Olfati-Saber, R. and Murray, R. (2004) Consensus problems in networks of agents with switching topology and time-delays. *IEEE Transactions on Automatic Control*, **49** (9), 1520–1533, doi:10.1109/TAC.2004.834113.

20 Xiao, L., Boyd, S., and Lall, S. (2005) A scheme for robust distributed sensor fusion based on average consensus. *Proceeding of the Fourth*

International Information Processing in Sensor Networks, Los Angeles, CA, pp. 63–70.

21 Olfati-Saber, R. and Shamma, J. (2005) Consensus filters for sensor networks and distributed sensor fusion, in *44th IEEE Conference on Decision and Control and 2005 European Control Conference*, pp. 6698–6703, doi:10.1109/CDC.2005.1583238.

22 Kar, S., Aldosari, S., and Moura, J. (2008) Topology for distributed inference on graphs. *IEEE Transactions on Signal Processing*, **56** (6), 2609–2613, doi:10.1109/TSP.2008.923536.

23 Kar, S. and Moura, J. (2008) Sensor networks with random links: Topology design for distributed consensus. *IEEE Transactions on Signal Processing*, **56** (7), 3315–3326, doi:10.1109/TSP.2008.920143.

24 Xiao, L. and Boyd, S. (2004) Fast linear iterations for distributed averaging. *Systems & Control Letters*, **53**, 65–78.

25 Olfati-Saber, R., Fax, J., and Murray, R. (2007) Consensus and cooperation in networked multi-agent systems. *Proceedings of the IEEE*, **95** (1), 215–233.

26 Dimakis, A., Kar, S., Moura, J., Rabbat, M., and Scaglione, A. (2010) Gossip algorithms for distributed signal processing. *Proceedings of the IEEE*, **98** (11), 1847–1864, doi:10.1109/JPROC.2010.2052531.

27 Ma, J., Zhao, G., and Li, Y. (2008) Soft combination and detection for cooperative spectrum sensing in cognitive radio networks. *IEEE Transactions on Wireless Communications*, **7** (11), 4502–4507, doi:10.1109/T-WC.2008.070941.

28 Quan, Z., Cui, S., and Sayed, A. (2008) Optimal linear cooperation for spectrum sensing in cognitive radio networks. *IEEE Journal of Selected Topics in Signal Processing*, **2** (1), 28–40, doi:10.1109/JSTSP.2007.914882.

29 Digham, F.F., Alouini, M.-S., and Simon, M.K. (2007) On the energy detection of unknown signals over fading channels. *IEEE Transactions on Communications*, **55**, 21–24.

30 Urkowitz, H. (1967) Energy detection of unknown deterministic signals. *Proceedings of the IEEE*, **55** (4), 523–531, doi:10.1109/PROC.1967.5573.

31 Zhang, W., Wang, Z., Guo, Y., Liu, H., Chen, Y., and Mitola III, J. (2011) Distributed cooperative spectrum sensing based on weighted average consensus, in *2011 IEEE Global Telecommunications Conference*, pp. 1–6, doi:10.1109/GLOCOM.2011.6134149.

32 Kay, S.M. (1993) *Fundamentals of Statistical Signal Processing: Estimation Theory*, Prentice-Hall, Inc., Upper Saddle River, NJ.

33 Kostylev, V. (2002) Energy detection of a signal with random amplitude, in *IEEE International Conference on Communications*, vol. 3, pp. 1606–1610, doi:10.1109/ICC.2002.997120.

34 Visser, F.E., Janssen, G.J., and Pawelczak, P. (2008) Multinode spectrum sensing based on energy detection for dynamic spectrum access, in. *IEEE Vehicular Technology Conference*, Singapore, pp. 1394–1398.

35 Meyer, C. (2000) *Matrix Analysis and Applied Linear Algebra*, SIAM, Philadelphia, PA.

36 Sardellitti, S., Giona, M., and Barbarossa, S. (2010) Fast distributed average consensus algorithms based on advection-diffusion processes. *IEEE Transactions on Signal Processing*, **58** (2), 826.

37 Kar, S., Aldosari, S., and Moura, J. (2008) Topology for distributed inference on graphs. *IEEE Transactions on Signal Processing*, **56** (6), 2609–2613, doi:10.1109/TSP.2008.923536.

38 Elsner, L., Koltracht, I., and Neumann, M. (1990) On the convergence of asynchronous paracontractions with applications to tomographic reconstruction from incomplete data. *Linear Algebra and Its Applications*, **130**, 65–82.

39 Meiss, J.D. (2007) *Differential Dynamical Systems*, SIAM Mathematical Modeling and Computation, Philadelphia, PA.

40 Jakovetic, D., Xavier, J., and Moura, J.M.F. (2010) Consensus in correlated random topologies: Weights for finite time horizon, in *IEEE International Conference on Acoustics Speech and Signal Processing*, pp. 2974–2977, doi:10.1109/ICASSP.2010.5496139.

41 Boyd, S., Ghosh, A., Prabhakar, B., and Shah, D. (2006) Randomized gossip algorithms. *IEEE Transactions on Information Theory*, **52** (6), 2508–2530, doi:10.1109/TIT.2006.874516.

42 Zhang, W., Wang, Z., Liu, H., Guo, Y., Chen, Y., and Mitola, J. (2015) Distributed consensus-based weight design for cooperative spectrum sensing. *IEEE Transactions on Parallel and Distributed Systems*, **26** (1), 54–64.

5

Distributed Consensus Filter for Radio Environment Mapping

5.1 Introduction

Radio environment mapping (REM) [1] mainly refers to an integrated database that provides multidomain environmental information and prior knowledge for cognitive radios, such as the geographical features, available services and networks, and locations and activities of neighboring radios. Among those, one of the fundamental problems is the estimation and dynamic tracking of the radio signal propagation map, such as power spectral density map estimation [2, 3], or as an alternative, the channel gain estimation [4] and tracking [5]. The REM studied in this chapter aims at using distributed sensors to recover dynamic radio signal spatial propagation and spectral energy distribution in a given frequency range. Recovered signal propagation maps facilitate dynamic spectrum sharing and help communicate the radio spectrum knowledge among common users of cognitive radio networks.

The REM estimation and tracking is challenging, as localizing transmitters may not be allowed by legislations. Without prior-known behaviors of the transmitters, the REM can be modeled as an uncertain dynamical system, such as spatial–temporal Gaussian random fields [6]. Existing REM results adopting this model include the radio tomographic imaging [7, 8] based on the medium-scale correlated shadow fading characterization, passive localization [9], and intruder detection [10] by analyzing the interaction between the signal propagation and the environmental geometry. However, most of the work on REM uses centralized methods, where a central data collection and processing machine is available to generate the global radio map. Those methods suffer the dependency of reporting channels, bandwidth constraints, and scalability issues [11]. There has been limited work on distributed solutions to the REM problem without a central station.

From the decentralized control and sensing perspective, distributed Kalman filters play an important role in sensor networks to estimate and recover uncertain dynamic fields, such as temperature over an area in the ocean [12]. By local neighboring communications, distributed sensors compute centralized global information in each Kalman filter iteration using consensus

Distributed Cooperative Control: Emerging Applications, First Edition. Yi Guo.
© 2017 John Wiley & Sons, Inc. Published 2017 by John Wiley & Sons, Inc.
Companion website: www.wiley.com/go/Guo/DistributedCooperativeControl

algorithm or distributed filter without a centralized fusion center. Based on this scheme, consensus-based distributed Kalman filters are developed for decentralized environment modeling in various applications. Lynch et al. developed the proportional-integral (PI) consensus filter [13], which adopts Kalman filter to model environments using mobile sensor networks [12]. To improve estimation accuracy in the environment represented by random fields, Cortes [14] developed distributed Kriged Kalman filter using PI consensus filter for spatial estimation. To reduce the computational cost introduced by the PI consensus filter, Saber proposed a suboptimal Kalman filter tracking algorithm [15] to facilitate algorithm implementation. However, existing work on distributed Kalman filter assumes that system parameters are either known or have been prior-estimated, which may not be feasible as prior knowledge is sometimes hard to estimate in practical systems, such as the REM problem discussed in this chapter.

For unknown system estimation, two main techniques commonly used in the literature, the maximum likelihood estimation [16] and the expectation–maximization (EM) [17], are both centralized methods. Inspired by the fact that there are no distributed solutions for environmental estimation and modeling with unknown system models, we study distributed consensus-based estimation and tracking of an uncertain field, and apply the method to solve the dynamic REM problem in the presence of uncertain system parameters.

In this chapter, we consider distributed consensus-based Kalman filtering to estimate and track dynamic REM with correlated shadowing. We adopt random fields to model the REM as an uncertain dynamical system, of which, the spatial and temporal dynamics are decoupled using function expansion. We develop consensus-based Kalman filtering with distributed EM algorithm to estimate the temporal dynamics and the system parameters that are used to recover and track the REM. Simulation results show the effectiveness of recovering the shadowing phenomenon and the dynamic tracking capability of REM.

The chapter contributes to both the fields of radio communication and distributed sensor networks. Specifically, the main contribution is twofold:

1. Comparing to existing work on REM where REM is constructed on a central station, we propose a distributed solution and consider shadowing effects of obstacles in the environment. The recovered REM provides both the energy distribution of a dynamic radio signal source and the environmental information (i.e., obstacle locations).
2. Comparing to existing work on consensus-based Kalman filter, we relax the assumption that system parameters are known a priori, and estimate unknown system parameters online, which is done by embedding the expectation maximization method in the Kalman filtering process. Consensus-based Kalman–EM filter is proposed accordingly.

Organization: The subsequent sections are organized as follows: In Section 5.2, we formulate the distributed REM tracking problem and illustrate the modeling process. Section 5.3 describes the distributed Kalman filtering techniques for REM tracking and how to estimate the unknown parameters. Section 5.4 analyzes our proposed algorithm. Section 5.5 shows simulation results. Finally in Section 5.6, we present the conclusion and future work.

5.2 Problem Formulation

5.2.1 System Configuration and Distributed Sensor Placement

As shown in Figure 5.1, we consider a dynamic radio transmitter moving in a bounded area with the size of x_w in length and y_w in width where obstacles exist. To explore the radio signal distribution in this area, radio frequency (RF) sensors are deployed for radio signal strength detection, and they can communicate with their neighboring sensors within their communication range R_c, using the communication topology indicated in Figure 5.2. In this chapter, we assume the following:

1. **The radio signal:** The radio transmitter moves in a bounded area, and the radio signal propagation model is represented in Appendix E.
2. **The environment:** There are stationary obstacles with unknown positions.

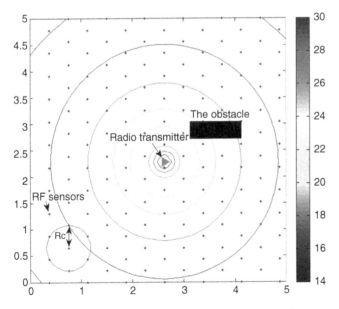

Figure 5.1 The REM system setup. The blue dots denote the RF sensors. The black rectangle represents the stationary obstacle. The solid triangle is the RF transmitter. The red circle represents the RF sensor communication range with the radius R_c. The color bar indicates the radio signal strength measurement value. (*See insert for color representation of the figure.*)

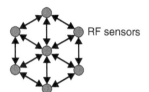 RF sensors

Figure 5.2 Sensor communication topology. The solid circles denote the RF sensors.

3. **RF sensors and placement:** The sensors are uniformly deployed and can communicate with their neighboring sensors within their communication range R_c. We assume each RF sensor can only detect its current position, signal strength measurement value, and the corresponding covariance of the measurement noise. They can also receive information from their one-hop neighboring sensors using the communication topology defined in Figure 5.2. In addition to the sensing capability, we assume the sensors have onboard computational power and can process measured data real time.

In this chapter, we define the global coordinates as follows: the positive x-axis pointing to the right horizontal direction, and the positive y-axis pointing to the up vertical direction. As shown in Figure 5.1, the sensor placement scheme is described in the following text. We place l columns parallel to the y-axis, and each column contains m sensors. The coordinate of the sensor locating at the eth row ($1 \le e \le m$) and fth column ($1 \le f \le l$) is at (x_c^{ef}, y_c^{ef}):

$$
(x_c^{ef}, y_c^{ef}) =
\begin{cases}
\left(\frac{\sqrt{3}}{2} R_c + \frac{\sqrt{3}}{2}(f-1)R_c, (e-1)R_c \right), \\
\quad \text{if } f \text{ is an odd integer;} \\
\left(\frac{\sqrt{3}}{2} R_c + \frac{\sqrt{3}}{2}(f-1)R_c, \frac{1}{2}R_c + (e-1)R_c \right), \\
\quad \text{if } f \text{ is an even integer.}
\end{cases}
$$

The number of sensors in the odd column m_{odd} and even column m_{even} are expressed as follows:

$$
\begin{cases}
m_{\text{odd}} = m_{\text{even}} + 1 = \text{Int}\left(\frac{y_w}{R_c} \right) + 1, \\
\quad \text{if } \text{Rem}(\frac{y_w}{R_c}) < \frac{R_c}{2}; \\
m_{\text{odd}} = m_{\text{even}} = \text{Int}\left(\frac{y_w}{R_c} \right) + 1, \\
\quad \text{if } \text{Rem}(\frac{y_w}{R_c}) \ge \frac{R_c}{2}.
\end{cases}
$$

Here, $\text{Int}(x)$ is the integer part of x, and $\text{Rem}(x) = x - \text{Int}(x)$. Also, the number of column l is $l = \text{Int}(\frac{x_w}{\frac{\sqrt{3}}{2} R_c})$.

Remark 5.1   Since we assume the sensor communication range as a circle, we adopt the sensor placement scheme as in Ref. [18]. For a more detailed analysis of coverage sensor placement, interested readers may refer to Ref. [19].

We are interested in distributed estimation and recovery of the REM caused by the dynamical RF signal. We formally define our problem in Section 5.2.2.

5.2.2 The Model and Problem Statement

The radio signal propagation model is described in Appendix E, where the signal strength for each location cell is explicitly computed according to (E.1). However, it is not practical to use (E.1) in constructing REM because it needs the prior–known trajectory of the transmitter to determine the temporal dynamics P_0 and L_a in (E.1), while localizing radio transmitter is not allowed by legislations. Also, without knowing the environment geometry and behaviors of the radio transmitter, the spatial–temporal dynamics of shadow fading, the S_a term in (E.1), cannot be determined. This is the fundamental reason that the radio propagation process is usually modeled in a different way.

Since the interaction between the signal and the environmental geometry evolves as uncertain spatio–temporal dynamics, it is common to model the radio signal propagation as random fields [4–7]. Specifically, dropping the subscript a in (E.1) for convenience, the radio signal y at any cell a and time k is regarded as a random field $y(k, s)$, where $s \in \mathbb{R}^2$ is the space coordinate and $k \in \mathbb{R}_+$ is the time index. We have the same assumption [4–7] that any pair of $y(k_1, s_1)$ and $y(k_2, s_2)$, $s_1, s_2 \in s$, $k_1, k_2 \in k$, only correlated in space not in time, that is,

$$\mathrm{Cov}(y(k_1, s_1), y(k_2, s_2)) = C(\|s_1 - s_2\|^2)\delta(k_1 - k_2), \tag{5.1}$$

where $C(\|s_1 - s_2\|^2)$ is a positive symmetric function and $\delta(k_1 - k_2)$ is the Dirac delta function. Neglecting the spatial correlated noise, the random field $y(k, s)$ from the Mercer's theorem has the function basis expansion as

$$y(k, s) = \sum_{j=1}^{\infty} \xi_j(k)\phi_j(s), \tag{5.2}$$

where $\xi(k)$ describes a temporal process consisting of infinite components, and $\phi(s)$ is the corresponding function basis to describe the spatial variation. For the consideration of computational cost, (5.2) usually takes the finite expansion as

$$y(k, s) = \sum_{j=1}^{M} \xi_j(k)\phi_j(s). \tag{5.3}$$

In the compact form, we have

$$y(k, s) = \phi(s)\xi(k), \tag{5.4}$$

where $\phi(s) = [\phi_1(s), ..., \phi_M(s)]$, and $\xi(k) = [\xi_1(k), ..., \xi_M(k)]^T$.

The major convenience from the expansion (5.2) is the decoupling of temporal dynamics and spatial variation of $y(k, s)$. For temporal dynamics, $\xi(k)$ is assumed to evolve as Ref. [20],

$$\xi(k) = A\xi(k-1) + v(k), \tag{5.5}$$

where A is the system matrix of temporal dynamics of the REM behaviors, $v(k)$ denotes the Gaussian noise with zero mean and covariance matrix $Q(k)$.

For a network consisting of n RF sensors, the measurement of the ith RF sensor at the corresponding cell is given by

$$Y_i(k,s) = y(k,s) + \zeta(k,s), \tag{5.6}$$

where $\zeta(k,s)$ denotes the Gaussian measurement noise with zero mean and covariance $R(k)$, which assumes to be not spatial correlated in this chapter for simplicity.

Considering the system model (5.4), (5.5), and (5.6), the Kalman filter technique can be used to estimate the system state ξ for recovering the signal strength y, which requires the system matrix A in (5.5) prior-known. However, due to uncertainties in practical systems, the system matrix A, is often time-varying and unknown so that the accurate estimation for ξ becomes challenge. Therefore, based on the model, the **distributed REM tracking problem** is defined as to *estimate and track $y(k,s), k \in R_+, s \in R^2$ using distributed sensor measurements (5.6) in the random Gaussian field with the unknown matrix A in the temporal dynamics (5.5)*.

Remark 5.2   The model (5.4)–(5.5) presented in this section is similar to the model in Ref. [12]. The major difference is that we assume that the system matrix A in (5.5) is unknown, while A is known or prior-estimated in Ref. [12]. A more detailed model for distributed tracking considering additive spatial correlated noise in a random field is presented in Ref. [14], where distributed Kriged Kalman filtering techniques was used to solve the problem.

5.3   Distributed REM Tracking

In this section, we present a distributed REM tracking algorithm to solve the distributed REM tracking problem defined in the above section. The main idea is to have each sensor distributively run a Kalman filter–based estimation algorithm. In the distributed configuration, each sensor can only measure its local receiving signal strength. However, global information propagates through the connected radio network as time elapses, and eventually a consensus global map is obtained by each individual sensor. This information diffusion process is achieved by employing the so-called consensus filtering technique, which is reviewed in Section 2.4.

Next, we first introduce the strategy to online estimate the system matrix A in (5.5) using an EM algorithm. We then present the algorithm combining Kalman filtering with the EM algorithm. Since the algorithm uses global information, we then describe our PI consensus filter method to distributively obtain the global sensing information.

5.3.1 System Matrix Estimation

In this section, we introduce our method to online estimate the unknown system matrix A in (5.5). Inspired by the Bayesian perspective of view, we utilize the EM algorithm by maximizing the expectation of log-likelihood, which fits the system matrix to the data of the system. This algorithm involves two steps: (i) maximize the expectation likelihood in terms of the system state using Kalman filter (E-step) and (ii) maximize the expectation likelihood in terms of the system matrix A (M-step). In our problem, for the system (5.4), (5.5), and (5.6), the estimated state $\xi(k)$ and observation $Y(k)$ follow the Gaussian process: $\xi(k) \sim N(A\xi(k-1), Q)$ and $Y(k) \sim N(\phi\xi(k), R)$, in which $(\cdot)(k)$ denotes the value of the variable at time k. The joint likelihood is:

$$\log P\left(\{\xi(1:\hat{k})\}, \{Y(1:\hat{k})\}|A(\hat{k})\right)$$

$$= \log\left[P(\xi(1)) \prod_{k=2}^{\hat{k}} P(\xi(k)|\xi(k-1)) \prod_{k=1}^{\hat{k}} P(Y(k)|\xi(k))\right],$$

$$(5.7)$$

where

$$P(Y(k)|\xi(k)) = \exp\left\{-\frac{1}{2}[Y(k) - \phi\xi(k)]^T R^{-1} \right.$$
$$\left. \cdot [Y(k) - \phi\xi(k)]\right\}(2\pi)^{-n/2}|R|^{-1/2},$$

$$P(\xi(k)|\xi(k-1)) = \exp\left\{-\frac{1}{2}[\xi(k) - A\xi(k-1)]^T Q^{-1} \right.$$
$$\left. \cdot [\xi(k) - A\xi(k-1)]\right\}(2\pi)^{-n/2}|Q|^{-1/2}$$

with observations $Y(1:\hat{k}) = (Y(1), ..., Y(\hat{k}))$ and states $\xi(1:\hat{k}) = (\xi(1), ..., \xi(\hat{k}))$.

At time $\hat{k}$, we perform the E-step of EM, requiring to compute the expected log-likelihood E_L, taking unobserved state ξ into account [21],

$$E_L(\hat{k}) = E[\log P(\xi(1:\hat{k}), Y(1:\hat{k}))|Y(1:\hat{k}); A(\hat{k})] \quad (5.8)$$

in which it requires to calculate

$$\hat{\xi}(k) = E[\xi(k)|Y(1:k); A(k)] \quad (5.9)$$
$$P(k) = E[\xi(k)\xi^T(k)|Y(1:k); A(k)] \quad (5.10)$$
$$P((k), (k-1)) = E[\xi(k)\xi^T(k-1)|Y(1:k); A(k)]. \quad (5.11)$$

The most accurate estimation of the system matrix $A(k)$ appears when (5.8) is maximized at time k, which is the M-step of EM algorithm. We obtain

$$\frac{\partial E_L(\hat{k})}{\partial A(\hat{k})} = -\sum_{k=1}^{\hat{k}} P((k), (k-1)) + \sum_{k=1}^{\hat{k}} Q^{-1}A(k)P(k-1) = 0, \quad (5.12)$$

which yields

$$A(\hat{k}) = \left(\sum_{k=1}^{\hat{k}} P((k), (k-1)) \right) \left(\sum_{k=1}^{\hat{k}} P(k-1) \right)^{-1}. \tag{5.13}$$

5.3.2 Kalman–EM Filter

The fundamental problem of REM tracking is to simultaneously estimate the unobserved system state ξ in (5.5) for each sensor, which is solved by Kalman filtering techniques. Since this process needs to use the system matrix A, we embed the EM algorithm introduced above into the Kalman filtering process, and present our Kalman–EM filter in this subsection.

Let $\tilde{\xi}$ represent the estimate of ξ, and V be the $M \times M$ covariance matrix correlated with estimate errors. Define the information matrix $\sigma(k) = V^{-1}(k)$, the information vector $\delta(k) = V^{-1}(k)\xi(k)$ and $M \times M$ scalar matrix $K_{M \times M}$. Based on initializations $A(0)$, $\sigma(0)$ and $\delta(0)$, both information vector τ and matrix σ at time k is obtained by the following steps of Kalman filter:

Prediction:

$$\tilde{\sigma}(k) = (A(k-1)\sigma^{-1}(k-1)A^T(k-1) + Q)^{-1} \tag{5.14}$$

$$\tilde{\delta}(k) = \tilde{\sigma}(k)A(k-1)\sigma^{-1}(k-1)\delta(k-1) \tag{5.15}$$

Correction:

$$\sigma(k) = \tilde{\sigma}(k) + \phi^T R^{-1}\phi \tag{5.16}$$

$$\delta(k) = \tilde{\delta}(k) + \phi^T R^{-1}Y(k) \tag{5.17}$$

Scalar matrix:

$$K(k) = (\tilde{\sigma}^{-1}(k) - \sigma^{-1}(k))\tilde{\sigma}(k). \tag{5.18}$$

Also, (5.10) and (5.11) have the solutions for (5.8) as follows:

$$P(k) = \hat{V}(k-1) + \hat{\xi}(k-1)\hat{\xi}^T(k-1) \tag{5.19}$$

$$P((k), (k-1)) = \hat{V}((k), (k-1)) + \hat{\xi}(k)\hat{\xi}^T(k-1). \tag{5.20}$$

Here, $\hat{V}(k-1)$, $\hat{\xi}(k-1)$ and $\hat{V}((k), (k-1))$ update all previous expectations of $\tilde{V}(k)$, $\tilde{\xi}(k)$ and $\tilde{V}((k), (k-1))$, respectively, for accurately estimating $A(k)$ once a new estimated state is generated, and $\tilde{V}((k), (k-1))$ denotes the cross-covariance matrix between two consecutive time instants. Specifically, this is the E-step of Kalman–EM filter that performs backward recursions for $k = \hat{k}, ..., 1$ on the equations,

$$J(k-1) = \sigma^{-1}(k-1)A^T(k-1)\tilde{\sigma}(k-1) \tag{5.21}$$

$$\hat{V}(k-1) = \sigma^{-1}(k-1)$$

$$\qquad + J(k-1)\left(\hat{V}(k) - \tilde{\sigma}^{-1}(k-1) \right)J^T(k-1) \tag{5.22}$$

$$\hat{\xi}(k-1) = \sigma^{-1}(k-1)\delta(k-1)$$

$$+ J(k-1)\left(\hat{\xi}(k) - A(k-1)\sigma^{-1}(k-1)\delta(k-1)\right) \tag{5.23}$$

which is initialized as $\hat{V}(k) = \sigma^{-1}(k)$ and $\hat{\xi}(k) = \sigma^{-1}(k)\delta(k)$.

For $k = \hat{k}, ..., 2$ on the equation, we have

$$\hat{V}((k-1), (k-2)) = \sigma^{-1}(k-1)J^T(k-2) + J(k-1)\left(\hat{V}((k), (k-1))\right.$$

$$\left. - A(k-1)\sigma^{-1}(k-1)\right)J^T(k-2) \tag{5.24}$$

which is initialized as $\hat{V}((k), (k-1)) = (I - K(k))A(k-1)\sigma^{-1}(k-1)$, where I is an identity matrix of dimension $M \times M$.

In the M-step, (5.10) and (5.11), which are obtained from the above backforward iteration, are then combined with (5.13), to obtain the updated system matrix $A(k)$, which serves as the outcome of the differentiation of (5.8).

5.3.3 PI Consensus Filter for Distributed Estimation and Tracking

In the Kalman–EM filter algorithm described above, we fuse information from *all* sensors to estimate REM in the presence of an unknown system matrix. Specifically, in (5.16) and (5.17), the terms $\phi^T R^{-1}\phi$ and $\phi^T R^{-1}Y(k)$ are global quantities, as

$$\phi^T R^{-1}\phi = \sum_{i=1}^n \phi(s_i)^T R_i^{-1}\phi(s_i), \tag{5.25}$$

$$\phi^T R^{-1}Y(k) = \sum_{i=1}^n \phi(s_i)^T R_i^{-1}Y_i(k), \tag{5.26}$$

where n is the number of RF sensors in the area. That is, in order to calculate $\phi^T R^{-1}\phi$ and $\phi^T R^{-1}Y(k)$, we need the global information for $i = 1, ..., n$. However, in our system setup, each sensor only knows its own parameters R_i and $Y_i(k)$ using its own position and measurement. As mentioned earlier, each sensor can communicate with its neighbors and eventually get the global information through information propagation by using the consensus filter technique. To achieve this, we have each sensor implement a PI consensus filter [12, 13]. The discrete consensus filter is given by

$$v_i(\tau) = v_i(\tau - 1) + \beta \left\{ \gamma[u_i - v_i(\tau - 1)] \right.$$

$$-K_P \sum_{j \in N_i} \left[v_i(\tau - 1) - v_j(\tau - 1)\right]$$

$$\left. + K_I \sum_{j \in N_i} [\eta_i(\tau - 1) - \eta_j(\tau - 1)] \right\}$$

$$\eta_i(\tau) = \eta_i(\tau - 1) - \beta \left\{ K_I \sum_{j \in N_i} \left[v_i(\tau - 1) - v_j(\tau - 1)\right] \right\} \tag{5.27}$$

where $j \in N_i$ denotes that sensor j is in the one-hop neighbor set, N_i, of sensor i, u_i is the sensor i's vector input to the sum, η_i and v_i are the sensor i's internal state, and the estimate of the average of all agents' inputs, respectively. $\gamma > 0$ is the parameter governing the rate at which new information replaces old information in the dynamic averaging process, and β is the step size for the PI consensus filter. K_p and K_I are estimator gains. For convenience, (5.27) can also be written in the following compact form:

$$
\begin{bmatrix} v(\tau) \\ \eta(\tau) \end{bmatrix} = \left(\begin{bmatrix} I & 0 \\ 0 & I \end{bmatrix} + \beta \begin{bmatrix} -\gamma I - L_p & L_I^T \\ -L_I & 0 \end{bmatrix} \right) \begin{bmatrix} v(\tau - 1) \\ \eta(\tau - 1) \end{bmatrix}
$$
$$
+ \begin{bmatrix} \gamma I \\ 0 \end{bmatrix} u. \tag{5.28}
$$

Here, I is the identity matrix, $L_p = K_p L$, $L_I = K_I L$, and L is the graph *Laplacian* defined in Section 2.2.

Suppose the total sensor number n is known to all sensors, let $u_i = n\phi(s_i)^T R_i^{-1} \phi(s_i)$, running the given protocol, $v_i, i = 1, \ldots, n$, converges to $\phi^T R^{-1} \phi$ as time elapses. Also, when $u_i = n\phi(s_i)^T R_i^{-1} Y_i(k)$, $v_i, i = 1, \ldots, n$, converges to $\phi^T R^{-1} Y(k)$. Therefore, *distributed* estimation is achieved using the PI consensus filter scheme shown earlier.

The procedure of our proposed algorithm is described in Algorithm 5.1. The ith sensor collects its radio energy measurement Y_i, position s_i (Line 1), calculates its local quantity $\phi(s_i)^T R_i^{-1} \phi(s_i)$ and $\phi(s_i)^T R_i^{-1} Y_i(k)$. (Line 3). After that, each of the sensors performs the PI consensus filtering toward calculating the global quantities $\phi^T R^{-1} \phi$ and $\phi^T R^{-1} Y(k)$ by their neighbor communication (Lines 4–10). Then they implement Kalman filter to estimate the system states (Line 12), run E-step backward recursion (Line 13), and estimate the parameter matrix A for M-step (Line 14), then update the dynamic REM by calculating (5.4) (Line 15). Thus, dynamic REM is recovered on each sensor real time.

Remark 5.3   In Algorithm 5.1, the PI consensus filter (39) is running in a fast time scale τ, and Kalman filter and EM backward recursions are running in a slow time scale k. Specifically, for each step of Kalman filter, the PI consensus filter converges to the global average of the individual inputs. We can tune the parameters of the PI consensus filter so that its convergence time is very fast.

5.4   Communication and Computation Complexity

We first introduce the following concepts and definitions:

1. B: the transition matrix defined in (5.28), which is given by

$$
B = \begin{bmatrix} I & 0 \\ 0 & I \end{bmatrix} + \beta \begin{bmatrix} -\gamma I - L_p & L_I^T \\ -L_I & 0 \end{bmatrix}.
$$

Algorithm 5.1 Distributed REM Tracking

Require:
Radio energy measurement Y, sensor position $\mathbf{s} = (\mathbf{m}, \mathbf{n})$.
Initialization: information vector $\delta(0)$, information matrix $\sigma(0)$, system matrix $A(0)$, state covariance matrix Q, and measurement noise covariance matrix R.

Ensure:
To enable each sensor to reach agreement on REM estimation, and track the dynamical REM.

1: **for** time step $k = 1$ to $\hat{k}$ **do**
2:    Each sensor i gets its corresponding position $s_i = (m, n)$ and sensor measurement of radio energy Y_i, then calculates its basis function $\phi(s_i)$.
3:    Each sensor calculates $\phi(s_i)^T R_i^{-1} \phi(s_i)$ and $\phi(s_i)^T R_i^{-1} Y_i(k)$.
4:    **for** $\tau = 1$ to p **do**
5:        Each sensor sets $u_i = n\phi(s_i)^T R_i^{-1} \phi(s_i)$,
6:        Each sensor gets its neighboring sensors' information $v_j, \eta_j, j \in N_i$,
7:        Each sensor runs PI consensus filter in (5.27) to get v_i and η_i;
8:        Each sensor sets $u_i = n\phi(s_i)^T R_i^{-1} Y_i(k)$,
9:        Each sensor repeats line 6 and 7,
10:   **end for**
11:   Each sensor returns consensused global quantities $v_i \rightarrow \phi^T R^{-1} \phi$ and $\rightarrow \phi^T R^{-1} Y(k)$.
12:   Each sensor runs Kalman filter line by line to calculate a priori, $\tilde{\sigma}(k)$ and $\tilde{\delta}(k)$, posteri $\sigma(k)$ and $\delta(k)$ and scalar matrix $K(k)$ using equations (5.14)–(5.18).
13:   Each sensor runs E-step backward recursion line by line to calculate $J(k-1)$, $\hat{V}(k-1)$, $\hat{\xi}(k-1)$ and $\hat{V}((k-1),(k-2))$ using equations of (5.21)–(5.24) .
14:   Each sensor substitutes $\hat{V}(k-1)$, $\hat{\xi}(k-1)$ and $\hat{V}((k-1),(k-2))$ into (5.19) and (5.20) to calculate $P((k),(k-1))$ and $P(k-1)$, and updates System Matrix $A(k)$ using equations of (5.13).
15:   Each sensor calculates the energy data $y(k, s)$ at all the points in the global map and recovers REM using (5.4).
16:   Next step $k + 1 \leftarrow k$.
17: **end for**

Let its eigenvalues denoted by λ_{Bi}, $i = 1, ..., 2n$. Let λ_{B2} be the second maximum value satisfying $|\lambda_{B2}| < 1$.

2. $\mu(\tau)$: the converging accuracy of PI consensus filter at time τ. It refers to the ratio of state converging and initial deviation, which is represented by

$$\mu(\tau) = \sup \frac{\left(v(\tau) - \frac{1}{n}\mathbf{1}\mathbf{1}^T u(0)\right)^T \left(v(\tau) - \frac{1}{n}\mathbf{1}\mathbf{1}^T u(0)\right)}{\left(v(0) - \frac{1}{n}\mathbf{1}\mathbf{1}^T u(0)\right)^T \left(v(0) - \frac{1}{n}\mathbf{1}\mathbf{1}^T u(0)\right)}.$$

where $v(\tau)$ is defined in (5.28), n is the sensor number defined in Section 5.3.3, and **1** denotes the vector with all elements one.

3. π: the recursion length of the EM algorithm.
4. M: the expansion number of function basis in (5.3).

Now we begin to analyze the communication and computation complexity for our proposed distributed REM tracking algorithm.

5.4.1 Communication Complexity

We focus on the communication complexity of PI consensus filter in Algorithm 5.1, because the communication only occurs when each sensor implements the PI consensus filter. According to Ref. [22], the communication complexity is described as the limit ratio between the average communication number per iteration and the minimum logarithmic upper bound of the converging accuracy, which is defined as

$$\Omega = \lim_{\mu_0 \to 0} \sup -\frac{E\left(\sum_{\tau=0}^{p} c_\tau\right)}{\log(\mu_0)}, \tag{5.29}$$

where c_τ is the number of communications per iteration. p is the iteration number of PI consensus filter satisfying $\mu(p) \leq \mu_0$ with $\mu_0 > 0$ meaning the upper bound of the converging accuracy.

Algorithm 5.1 adopts the sensor topology in Figures 5.1 and 5.2, which is an example of the limited communication scenario: each sensor only communicates to its six neighbors excluding boundary sensors. In this case, (5.29) writes

$$\Omega = -\frac{6n - 4(l + m) + 2}{2\log(|\lambda_{B2}|)}, \tag{5.30}$$

where l and m are defined in Section 5.2.1, respectively. Hence, the communication complexity is $O(n)$. If Algorithm 5.1 implements on the all-to-all communication case, the communication complexity is $O(n^2)$, according to (5.29).

5.4.2 Computation Complexity

Computation complexity is a measurement on the required number of steps to solve instances of a problem [23]. In Algorithm 5.1, we analyze the computation complexity on the following parts: PI consensus filter, Kalman filter, and EM algorithm. For the PI consensus filter, each sensor processes data from at most six neighbors, and takes at most p iterations for each REM tracking step. Thus, the computation complexity for the n sensors is $O(npM^2)$. The computation complexity for the Kalman filter is $O(nM^3)$ because the computations for each sensor involve a fixed number of matrix multiplications and inversions, whose computation complexities are both $O(M^3)$. In EM algorithm, each sensor requires the backward recursion whose maximum length is π, in which, a fixed number of matrix multiplications and inversions are also contained. Hence, the computation complexity for EM algorithm is $O(n\pi M^3)$. Combining the three cases, the computation complexity at each REM tracking step is $O(npM^2 + n\pi M^3)$. If Algorithm 5.1 utilizes the all-to-all communication topology, the computation complexity for each REM step is $O(pn^2 M^2 + \pi n^2 M^3)$.

5.5 Simulations and Performance Evaluation

In this section, we use MATLAB to perform simulation validation of our proposed algorithm on a two-dimensional (2D) bounded area. Following the system setup in Section 5.2.1, we generate REMs in both scenarios of dynamic and stationary radio transmitters according to the radio signal propagation model described in Appendix E. We implement our proposed algorithm, and simulate distributed sensors to estimate REMs in both scenarios for the dynamic and stationary radio transmitters. The performance are compared with existing methods.

5.5.1 Dynamic Radio Transmitter

We generate the dynamical REM based on the system setup of Section 5.2.1, where the bounded area consists of 47×41 regular cells, and the RF transmitter is moving around the obstacle. In the simulation, we denote the RF transmitter power in the form of the channel gain (dB). According to the radio propagation model (E.1), we set $P_0 = 20$ dB; the attenuation of the bth cell $x_b = 5$ if the cell is occupied by an obstacle, while $x_b = 0.1$ if the cell is in free space; the environmental noise $\iota_a(k)$ for each cell follows the Gaussian distribution: $\iota_a(k) \sim N(0, 1)$; the parameter $\Gamma = 0.1$; we ignore the multipath and interfere effect so that $F_a(k) = 0$. Figure 5.3 shows the energy distribution of RF transmitter at different time instants, in which the reddish area having the highest signal strength corresponds to the location of the radio transmitter, and the black rectangle denotes the obstacle.

In the simulation, $n = 156$ sensors with the communication range $R_c = 0.433$ over the bounded area as shown in Figure 5.1, our goal is to estimate and track the REM using Algorithm 5.1. To implement our proposed algorithm, each sensor sets the parameters as following: the function basis subset in (5.4) contains $M = 15$ component sets as

$$\phi(s_x, s_y) = [1, s_x, s_y, s_x^2, s_y^2, s_x s_y, s_x^3, s_y^3, s_x s_y^2, s_x^2 s_y, s_x^4, s_y^4, s_x s_y^3, s_x^2 s_y^2, s_x^3 s_y]$$

with s_x and s_y being the horizontal and vertical positions, respectively; the covariance matrix of the temporal dynamic noise in (5.5) is $Q = 0.01 I_{15 \times 15}$; the measurement covariance matrix for each sensor in (5.6) is $R = 0.1 I_{15 \times 15}$; each sensor initializes the Kalman filter parameters in (5.14) and (5.15) as $\delta_{(0)} = 0_{15 \times 1}$ and $\sigma_{(0)} = 0.0001 I_{15 \times 15}$; the parameters of PI consensus filter (5.27)

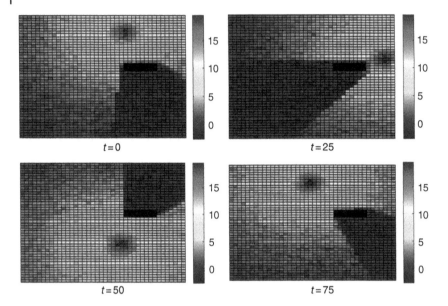

Figure 5.3 REM of the dynamic radio transmitter generated using (E.1) in Appendix E. The unit of time *t* is the second. (*See insert for color representation of the figure.*)

are $\gamma = 1.6$, $\beta = 0.5$, $K_p = 0.1$, and $K_I = 0.01$; the initializations of PI consensus filter are $v_i(0) = 0_{15\times15}$ and $\eta_i(0) = 0_{15\times15}$ for C estimation, and $v_i(0) = 0_{15\times1}$ and $\eta_i(0) = 0_{15\times1}$ for D estimation, and we randomly initialize the system matrix A.

From Figure 5.4, it can be seen that the initial REM is poorly recovered. However, with the tracking convergence by the distributed Kalman filter, the dynamical changing REM is tracked successfully, and the radio transmitter can be approximately localized, where the radio signal strength becomes high. Moreover, the radio signal distribution is simultaneously recovered, enabling the radio signal shadowing generated by the obstacle to be accurately displayed.

Comparison with Existing Distributed Method: We compare the REM tracking performance of our proposed algorithm with the existing method, which uses the distributed Kalman filter based on the PI consensus filter with a randomly chosen system matrix A [12]. We utilize the mean absolute error (MAE) as an evaluation criteria for REM recovery performance. The MAE is defined as

$$MAE = \frac{1}{\mathcal{N}} \sum_{a=1}^{\mathcal{N}} |\hat{y}_a(s) - y_a(s)|, \tag{5.31}$$

where $\mathcal{N}$ is the total cell number in the REM, and $\hat{y}_a(s)$ is the estimation of $y_a(s)$ in (E.1).

Now we compare our proposed Algorithm 5.1 with the existing method [12] that uses the same system and parameter setup. Figure 5.5 shows the

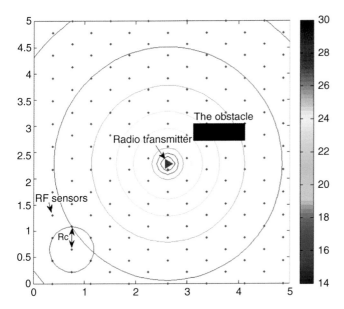

Figure 5.1 The REM system setup. The blue dots denote the RF sensors. The black rectangle represents the stationary obstacle. The solid triangle is the RF transmitter. The red circle represents the RF sensor communication range with the radius R_c. The color bar indicates the radio signal strength measurement value.

Distributed Cooperative Control: Emerging Applications, First Edition. Yi Guo.
© 2017 John Wiley & Sons, Inc. Published 2017 by John Wiley & Sons, Inc.
Companion website: www.wiley.com/go/Guo/DistributedCooperativeControl

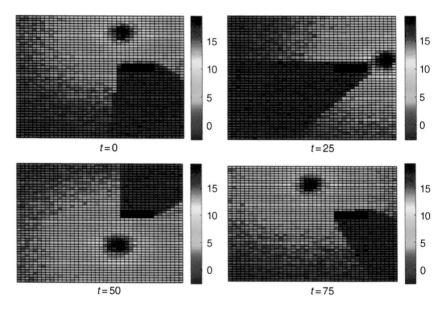

Figure 5.3 REM of the dynamic radio transmitter generated using (E.1) in Appendix E. The unit of time t is the second.

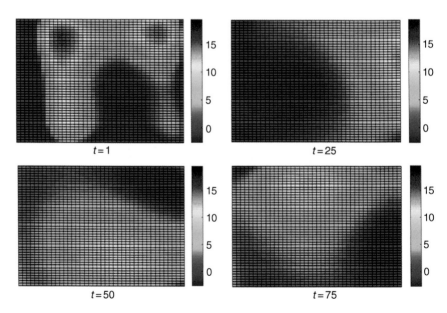

Figure 5.4 Distributed REM tracking in the case of dynamic radio transmitter using our proposed algorithm.

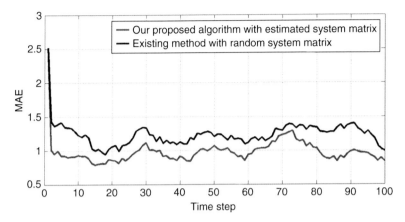

Figure 5.5 MAE comparison of our proposed algorithm with the existing distributed method in the case of dynamic radio transmitter.

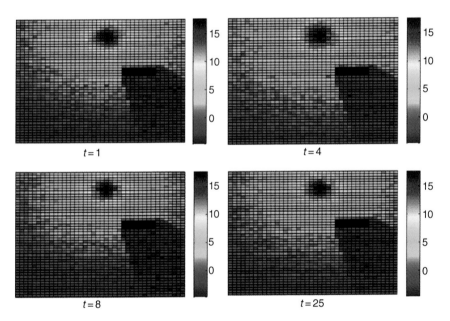

Figure 5.6 REM of the stationary radio transmitter generated using (E.1) in Appendix E. The unit of time t is the second.

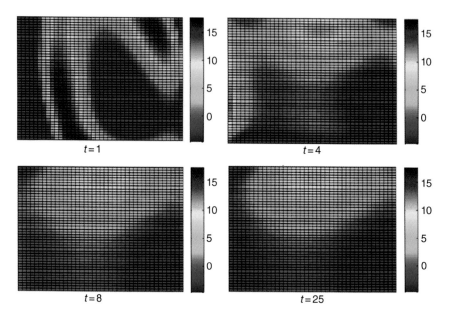

Figure 5.7 Distributed REM estimation in the case of stationary radio transmitter using our proposed algorithm.

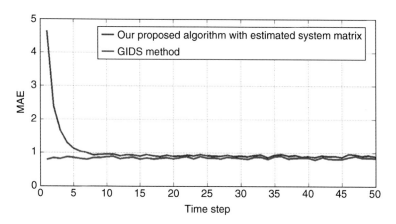

Figure 5.8 MAE comparison of our proposed algorithm with the exiting centralized methods in the case of stationary radio transmitter.

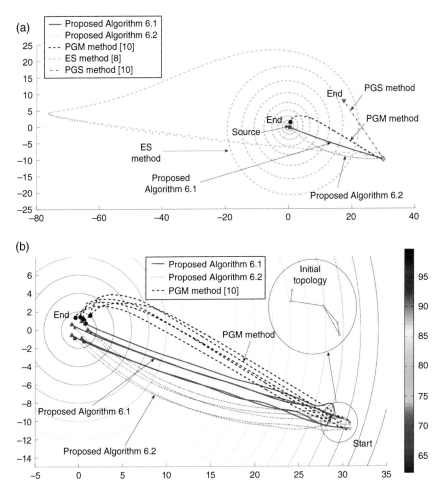

Figure 6.1 Simulation comparisons between PGS method [10], ES method [8], PGM method [10], and our Algorithms 6.1 and 6.2. The color bar denotes the concentration value of the source. (a) Trajectory of formation centers (for single robot case, it is the robot's own trajectory). The yellow diamond represents the start position. The red square is the end position of the robot by ES method and the proposed Algorithms 6.1 and 6.2. It is also the source position. The black dot and the pink triangle are the end position of the robot by PGM method and that of the robot by PGS method. (b) Trajectories of each robot for methods using multiple robots. The black dots, the blue triangles and the red squares are the end positions of robots by PGM method and Algorithms 6.1 and 6.2, respectively. The green dots represents the start positions. The contour of the field $p(\mathbf{x})$ is plotted in the figure.

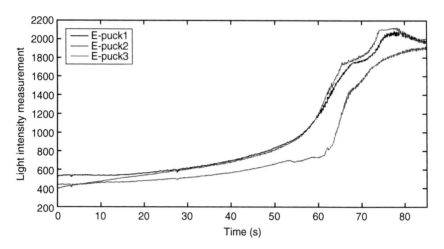

Figure 6.7 Robots' light intensity measurements in the all-to-all communication case.

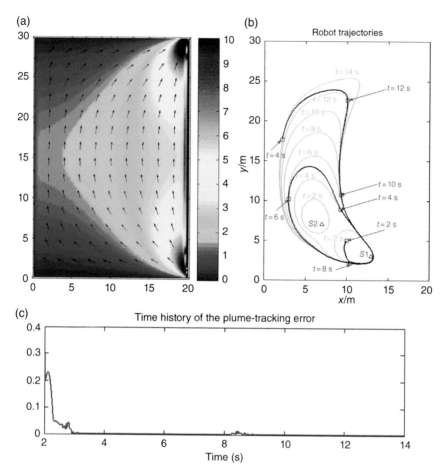

Figure 7.5 (a) The velocity field considered in the simulation, where the pseudo-color and the yellow arrow indicate the strength and the direction of the flow, respectively. (b) The robot trajectory for single-robot plume front tracking and patrolling, where S_1 and S_2 are the two sources; the green curves and the red squares represent the plume front contour and the robot positions, respectively; and black curve represents the robot trajectory. (c) The plume front tracking error.

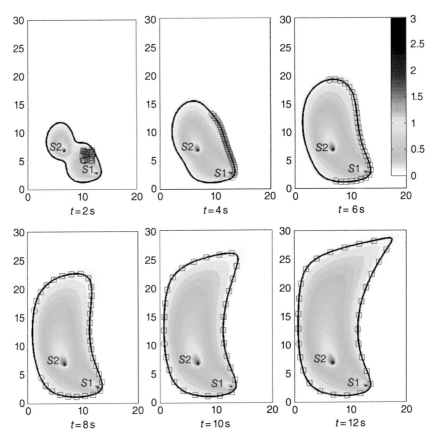

Figure 7.6 The snapshots of the robot movements for a typical simulation run with 30 robots for cooperative plume front tracking, where the pseudo-color indicates the concentration distribution at each time step with the scale shown in the colorbar aside; the black curve is the contour of the plume front with a concentration value $c_0 = 0.01$; the hollow square in blue and the hollow circle in red represent the position of the follower robots and the true position of the boundary robots, respectively; and the positions marked with S1 and S2 are the two sources starting propagating chemicals in the flow field shown in Figure 7.5 at time $t = 0$.

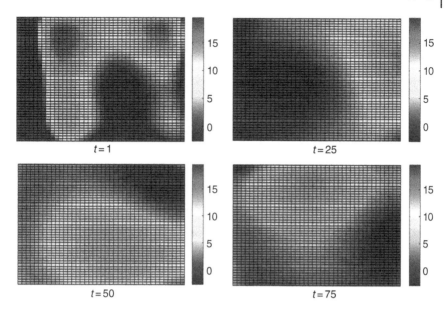

Figure 5.4 Distributed REM tracking in the case of dynamic radio transmitter using our proposed algorithm. (*See insert for color representation of the figure.*)

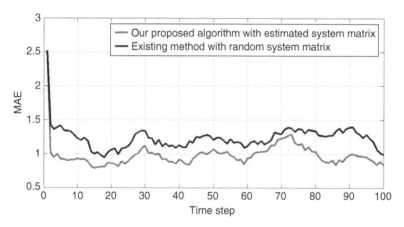

Figure 5.5 MAE comparison of our proposed algorithm with the existing distributed method in the case of dynamic radio transmitter. (*See insert for color representation of the figure.*)

comparison results. The high initial MAEs for both algorithms result from the random system initializations. Then, the MAEs of both algorithms decrease because of the convergence of the distributed Kalman filter. It is easy to see that our proposed algorithm which online estimates the system matrix A has lower MAE, and outperforms the existing algorithm implemented in Ref. [12]

that uses the random system matrix A. The MAE curve oscillates due to the radio transmitter dynamics.

5.5.2 Stationary Radio Transmitter

We generate the REM of the static radio transmitter locating at the randomly chosen position (2.63, 4.24). Other system setup is the same as in Section 5.5.1. Figure 5.6 shows the REM generated by the stationary radio transmitter at different time instants. Due to the stationary radio transmitter, the REM dynamics only depends on the time varying noises.

We use the same parameters as in Section 5.5.1, and still randomly initialize the system matrix A. Figure 5.7 shows the REM recovery results using our proposed algorithm. The REM at the beginning is not recovered because of the arbitrary parameter initializations. As the distributed Kalman filter in the algorithm converges, the signal strength becomes gradually better although the REM is not fully recovered yet, as seen from Figure 5.7 at $t = 4$ s. The fully recovered REMs are shown in Figure 5.7 at $t = 8$ s and $t = 25$ s.

5.5.3 Comparison with Existing Centralized Methods

Here, we first introduce the existing centralized REM methods in the scenario of the stationary radio transmitter, and then compare the REM tracking performance of our proposed algorithm with the existing methods.

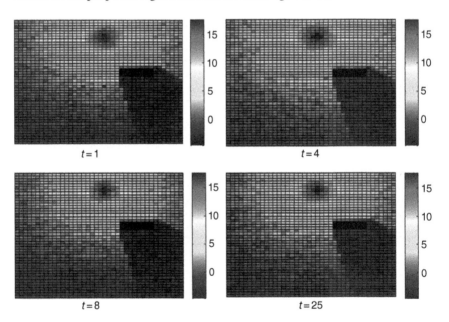

Figure 5.6 REM of the stationary radio transmitter generated using (E.1) in Appendix E. The unit of time *t* is the second. (*See insert for color representation of the figure.*)

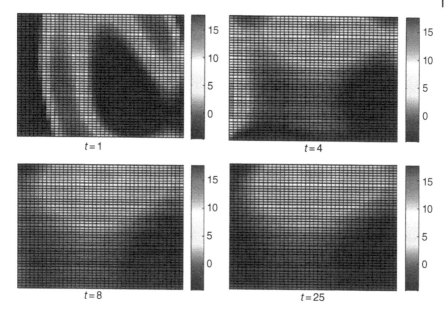

Figure 5.7 Distributed REM estimation in the case of stationary radio transmitter using our proposed algorithm. (*See insert for color representation of the figure.*)

Existing Centralized REM Methods: In the case of the REM generated by the stationary radio transmitter, a number of the existing methods using classical spatial interpolation techniques provide robust and reliable REM estimation from sensor measurements [24]. These methods include (i) the inverse distance weighting based modified Shepard's method (IDW-MSM) [25] computes the interpolants using the distance-based weighted Nodal function of the measurements within the predefined radius; (ii) the Kriging method [26] regards the spatial correlation–based weighted measurements as the interpolants with the consideration of random field model; and (iii) the Gradient plus inverse distance squared (GIDS) method [27] determines the interpolants by calculating the distance square–based weighted Talyor expansion of the measurements where the gradients are obtained using least-square technique. In Ref. [28], the REM recovery errors of these methods are compared using the centralized scheme, in which, the REM recovery is implemented on a central station. From the comparison, we can see that the REM recovery error decreases as the sensor number increases regardless of both indoor and ourdoor scenarios. Moreover, despite minor differences, the REM recovery errors of these three methods approach to each other, especially when the sensor number is relatively large. For this reason, we choose the GIDS method [27] as a representative of existing centralized methods.

Performance Comparison: We compare our proposed Algorithm 5.1 with the GIDS method [27]. For fair comparison, we use the same system and

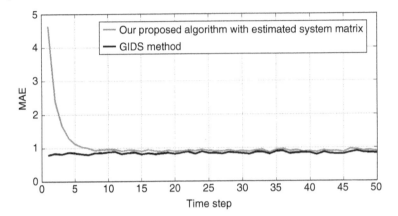

Figure 5.8 MAE comparison of our proposed algorithm with the exiting centralized methods in the case of stationary radio transmitter. (*See insert for color representation of the figure.*)

parameter setup as in Section 5.5.1. For the GIDS method, we choose the parameter n defined in Ref. [27] as $n = 156$, and the algorithm is implemented in the centralized scheme by sensor measurements gathered from the central station. Figure 5.8 shows the comparison results. It can be seen that the MAE of our proposed algorithm initially becomes high for the arbitrary parameter setups. In contrast, the MAE of the GIDS method is low since the centralized GIDS method requires only one step calculation on the central station, and the MAE represents the estimation error. As time elapses, the MAE of our proposed algorithm decreases as the result of convergence of the distributed Kalman filter implemented in our algorithm. Moreover, with the aid of the system matrix estimation, the tracking result gets better, and the final steady MAE representing estimation error is comparable to the centralized GIDS method.

5.6 Conclusion

In this chapter, we study distributed estimation and tracking of REM, which is modeled by radio signal propagation model as uncertain Gaussian random fields. Without localizing the radio transmitter, we develop distributed consensus-based Kalman–EM filtering techniques to estimate the unknown system parameters and to generated the REM heat map. We implement the method and show simulation results for both cases of stationary and dynamic radio transmitters. Simulation results demonstrate satisfactory performances of the proposed method, where spatial correlated shadowing effects are clearly recovered.

Notes

The results in this chapter were first published in Ref. [29].

Acknowledgment is given to ©2014 IEEE. Reprinted, with permission, from Kong, R., Zhang, W., and Guo, Y. (2014) Distributed estimation and tracking for radio environment mapping, in American Control Conference, pp. 464–470.

References

1 Zhao, Y. (2007) *Enabling cognitive radios through radio environment maps*, Ph.D. thesis, Virginia Polytechnic Institute and State University.

2 Bazerque, J. and Giannakis, G. (2010) Distributed spectrum sensing for cognitive radio networks by exploiting sparsity. *IEEE Transaction on Signal Processing*, **58** (3), 1847–1862.

3 Zhang, W., Wang, Z., Guo, Y., Liu, H., Chen, Y., and Mitola-III, J. (2011) Distributed cooperative spectrum sensing based on weighted average consensus, in *Proceedings of the IEEE Global Telecommunications Conference (GLOBECOM)*, Houston, TX, pp. 1–6.

4 Kim, S.J., Dall'Anese, E., and Giannakis, G. (2011) Cooperative spectrum sensing for cognitive radios using Kriged Kalman filtering. *IEEE Journal of Selected Topics in Signal Processing*, **5** (1), 24–36.

5 Dall'Anese, E., Kim, S.J., and Giannakis, G. (2011) Channel gain map tracking via distributed kriging. *IEEE Transactions on Vehicular Technology*, **60** (3), 1205–1211.

6 Riihijarvi, J., Mahonen, P., Wellens, M., and Gordzie, M. (2008) Characterization and modelling of spectrum for dynamic spectrum access with spatial statistics and random fields, in *Proceedings of the 19th IEEE International Symposium on Personal, Indoor and Mobile Radio Communications*, Cannes, pp. 1–6.

7 Patwari, N. and Agrawal, P. (2008) Effects of correlated shadowing: Connectivity, localization, and RF tomography, in *Proceedings of International Conference on Information Processing in Sensor Networks*, St. Louis, MO, pp.82–93.

8 Agrawal, P. and Patwar, N. (2009) Correlated link shadow fading in multi-hop wireless networks. *IEEE Transactions on Wireless Communications*, **8** (8), 4024–4036.

9 Wagner, B., Patwari, N., and Timmermann, D. (2012) Passive RFID tomographic imaging for device-free user localization, in *Proceedings of the 9th Workshop on Positioning Navigation and Communication*, Dresden pp. 120–125.

10 Wilson, J. and Patwari, N. (2010) Radio tomographic imaging with wireless networks. *IEEE Transactions on Mobile Computing*, **9** (5), 621–632.

11 Akyildiz, I.F., Lo, B.F., and Balakrishnan, R. (2011) Cooperative spectrum sensing in cognitive radio networks: A survey. *Physical Communication*, **4**, 40–62.

12 Lynch, K.M., Schwartz, I.B., Yang, P., and Freeman, R.A. (2008) Decentralized environmental modeling by mobile sensor networks. *IEEE Transaction on Robotics*, **24** (3), 710–724.

13 Freeman, R., Yang, P., and Lynch, K. (2006) Stability and convergence properties of dynamic average consensus estimators, in *Proceedings of the 45th IEEE Conference on Decision and Control*, San Diego, CA, pp. 338–343.

14 Cortes, J. (2009) Distributed kriged kalman filter for spatial estimation. *IEEE Transaction on Automatic Control*, **54** (12), 2816–2827.

15 Olfati-Saber, R. (2009) Kalman-consensus filter : Optimality, stability, and performance, in *Proceedings of the 48th IEEE Conference on Decision and Control*, Shanghai, China, pp. 7036–7042.

16 Gupta, N. and Mehra, R. (1974) Computational aspects of maximum likelihood estimation and reduction in sensitivity function calculations. *IEEE Transaction on Automatic Control*, **19** (6), 774–783.

17 Shumway, R. and Stoffer, D. (1982) An approach to time series smoothing and forecasting using the em algorithm. *Journal of Time Series Analysis*, **3** (4), 253–264.

18 Guo, Y. and Balakrishnan, M. (2006) Complete coverage control for non-holonomic mobile robots in dynamic environments, in *Proceedings of IEEE International Conference on Robotics and Automation*, Orlando, FL, pp. 1704–1709.

19 Bai, X., Xuan, D., Yun, Z., Lai, T.H., and Jia, W. (2008) Complete optimal deployment patterns for full-coverage and k-connectivity (k≤6) wireless sensor networks, in *Proceedings of the 9th ACM International Symposium on Mobile Ad Hoc Networking and Computing (MobiHoc)*, New York, NY, pp. 401–410.

20 Mardia, K., Goodall, C., Redfern, E., and Alonso, F. (1998) The Kriged Kalman filter. *Test*, **7** (2), 217–282.

21 Ghahramani, Z. and Hinton, G. (1996) Parameter estimation for linear dynamical systems, *Technical Report CRG-TR-96-2*, University of Toronto.

22 Mo, Y. and Slinopoli, B. (2010) Communication complexity and energy efficient consensus algorithm, in *Proceedings of the 2nd IFAC Workshop on Distributed Estimation and Control in Networked Systems*, Annecy, France, pp. 209–214.

23 Sipser, M. (2012) *Introduction to the Theory of Computation*, 3rd edn., Thomson Course Technology, Boston, MA.

24 Denkovski, D., Atanasovski, V., Gavrilovska, L., Riihijarvi, J., and Mahonen, P. (2012) Reliability of a radio environment map: Case of spatial interpolation techniques, in *Proceedings of the 7th International ICST Conference on Cognitive Radio Oriented Wireless Networks and Communications*, Stockholm, Sweden, pp. 248–253.

25 Renka, R.J. (1988) Multivariate interpolation of large sets of scattered data. *ACM Transactions on Mathematical Software*, **14** (2), 139–148.

26 Cressie, N.A.C. (1993) *Statistics for Spatial Data*, John Wiley & Sons., Inc., New York.

27 Nalder, I.A. and Wein, R.W. (1998) Spatial interpolation of climatic normals: Tests of a new methods in Canadian boreal forests. *Agricultural and Forest Meteorology*, **92** (4), 211–225.

28 Angjelicinoski, M., Atanasovski, V., and Gavrilovska, L. (2011) Comparative analysis of spatial interpolation methods for creating radio environment maps, in *Proceedings of the 19th Telecommunications Forum*, Belgrade, Serbia, pp. 334–337.

29 Kong, R., Zhang, W., and Guo, Y. (2014) Distributed estimation and tracking for radio environment mapping, in *American Control Conference*, pp. 464–470, doi:10.1109/ACC.2014.6859200.

Part II

Distributed Cooperative Control for Multirobotic Systems

Part II presents cooperative control techniques applied to distributed robotic systems. As reviewed in Section 1.2.2, robotics has been an important application area for recent development in cooperative control of multiagent systems. In particular, based on the consensus concept, formation control and its variations, such as leader-following control, vehicle platoon, flocking, and rendezvous, have been extensively studied for autonomous vehicles. In real-world applications of multirobotic systems, higher-level control objectives are needed for group tasks, such as search and rescue and source seeking, and formation control provides motion patterns of group members for highly efficient sensing and/or actuation. Chapters 6 and 7 in this part present real-world robot applications, where a group of robots are deployed to seek a source (e.g., chemical, gas, light, temperature, or radio signal sources), and to track the boundary of an underwater pollution plume, respectively. These two applications represent emerging needs for advanced robotics techniques to solve high-impact societal problems such as environmental monitoring.

Chapter 6 discusses source seeking by a group of mobile robots that have sensors to detect the field. The robots cooperatively estimate the gradient of the field concentration, and move gradient-ascending to the source in formation. To achieve distributed cooperative estimation, that is, to estimate the gradient with limited neighboring communication among the robots, a consensus filter is implemented and embedded in the least-square estimator (LS); thus, global quantities are obtained in consensus through information diffusion of the network. A light source-seeking experiment using real robot platforms is performed to demonstrate the performance of the distributed robot system.

Chapter 7 addresses a pressing need for robotics techniques to monitor environments, motivated by the recent Deepwater Horizon oil spill occurred in the Gulf of Mexico of the United States. Different from the source-seeking problem discussed in Chapter 6, a cooperative plume-tracking problem is defined as to track the dynamic prorogation of the boundary of the plume

Distributed Cooperative Control: Emerging Applications, First Edition. Yi Guo.
© 2017 John Wiley & Sons, Inc. Published 2017 by John Wiley & Sons, Inc.
Companion website: www.wiley.com/go/Guo/DistributedCooperativeControl

front by a group of mobile robots in a nearest-neighbor communication structure. The challenge is to estimate the plume front's trajectory based on the advection–diffusion equation of point-source dispersion. Formation control is also embedded in the tracking control law to have the multirobot team spread out along the plume front to achieve best sensing coverage. Simulation results show satisfactory performances.

6

Distributed Source Seeking by Cooperative Robots

6.1 Introduction

Source seeking has been studied in robots, whose potential applications include source localization of oil spill [1], chemical plume tracing [2], and cooperative foraging [3]. For a robot with the ability of measuring concentration gradients, a simple moving strategy by following gradient-ascending direction can complete the task. However, in practice, most robots are only equipped with sensors for concentration measurement instead of the gradient measurement. Similar problems to source seeking can be found in nature. For example, a male moth is able to approach a female one from far away by tracing the pheromone plume [2]. For a swarm of bacteria, without the ability of concentration gradient measurements, they still are able to find the source of beneficial chemicals [4]. Such phenomena give insight into the problem and inspire many studies in the field.

Various methods have been proposed to solve the distributed source-seeking problem, which include behavior-based source seeking [5–7] and control-based source seeking [8–11]. The behavior-based source seeking method often defines a set of elementary behaviors and a set of behavior combination rules. Different combinations of the elementary behaviors are activated in schedule to steer a single robot or a group of robots. The authors in Refs. [5, 6] design behavior-based source-seeking algorithms for a single robot. In Ref. [7], the method is extended to the scenarios with a group of robots. However, the inherent dynamics of the robot are often ignored for behavior-based source seeking, and there is no guarantee to reach the source eventually. In contrast, the control-based source-seeking method directly takes the robot dynamics into account to develop a control algorithm. Zhang et al. in Ref. [8] use extremum-seeking control theories to design a source-seeking control. In Ref. [9], the authors give a control law by combining a potential field control law and a gradient control law. In Ref. [10], two strategies for source seeking are considered. The first strategy uses one robot to perform the task with the historical data for estimating the gradient along the trajectory, while the second one uses a group of robots with projected gradient estimation. Ogren et al. [11] solve the problem by decoupling it into formation maintenance and

Distributed Cooperative Control: Emerging Applications, First Edition. Yi Guo.
© 2017 John Wiley & Sons, Inc. Published 2017 by John Wiley & Sons, Inc.
Companion website: www.wiley.com/go/Guo/DistributedCooperativeControl

leader following. Among existing literatures using control-based method for multiple robot source seeking, most of them ignore the gradient estimation error to ease the treatment, and the control algorithm often requires shared information with all robots in the group. These limitations inspire our study to design control algorithms that are robust to the gradient estimation error and also relax the communication requirements.

In this chapter, we use distributed robots for source seeking. The problem is modeled as a cooperative estimation and control problem. We first propose a source seeking algorithm with all-to-all communications. Theoretical analysis are given to prove that this algorithm enables the group of robots to approach the source. Then, a modified algorithm, which is fully distributed with limited communications among robots, is presented to solve the same source-seeking problem. Both the all-to-all and the limited communication algorithms are validated by simulations and real robot experiments.

The contribution of this chapter is twofold. First, we use the gradient estimation to guide the movement, and we provide a theoretical upper bound on the tracking error. Second, among existing methods using a group of robots, such as Refs. [9–11], each robot needs global information from all the others. In contrast, the algorithm with limited communications presented in this chapter is a fully distributed and scalable one, that is, each robot only needs to communicate with one-hop neighbors.

Organization: The rest of the chapter is organized as follows: Section 6.2 gives assumptions and formulations. Section 6.3 presents the first control algorithm with all-to-all communications. The second algorithm requiring only limited communications is proposed in Section 6.4. In Section 6.5, MATLAB simulations are performed to show the effectiveness of the proposed algorithms and their comparison with existing ones. In Section 6.6, experiments are conducted on a multi-robot testbed. Section 6.7 concludes the chapter.

6.2 Problem Formulation

Following the same model as in Refs. [8, 10, 11], we assume that the robot's motion is described by a double integrator:

$$\dot{\mathbf{x}}_i = \mathbf{v}_i$$
$$\dot{\mathbf{v}}_i = \mathbf{u}_i \text{ for } i = 1, 2, ..., n, \tag{6.1}$$

where $\mathbf{x}_i \in \mathbb{R}^k$, $\mathbf{v}_i \in \mathbb{R}^k$ and $\mathbf{u}_i \in \mathbb{R}^k$ are the position, velocity, and control input (acceleration) of the ith robot in a k-dimensional workspace.

We make the following assumptions on the environments:

Assumption 6.1 The scalar-valued distribution $p(\mathbf{x}) : \mathbb{R}^k \to \mathbb{R}$ is a concave function with respect to $\mathbf{x}$ and reaches its maximum at $\mathbf{x} = \mathbf{x}_s$.

Assumption 6.2   The Hessian matrix of $p(\mathbf{x})$ satisfies $-\xi_2 \le \lambda_{min}(\mathbf{H})$ and $\lambda_{max}(\mathbf{H}) \le -\xi_1$ for all $\mathbf{x}$ in the domain with $\xi_2 > \xi_1 > 0$.

Note that the Hessian matrix of $p(\mathbf{x})$, $\mathbf{H}(\mathbf{x})$, is an $n \times n$ square matrix with the ith row and jth column element defined as $(\partial^2 p / \partial x_i \partial x_j)$ for $\mathbf{x} = [x_1, x_2, \dots, x_n]^T$.

Remark 6.1   For the field $p(\mathbf{x})$ formed by a single source, such as a temperature field and an electric field, it has the maximum value at the source position $\mathbf{x}_s$ and reduces with the increase of distance from it. Assumption 6.1 is a simplification to this observation. It is also a commonly made assumption in optimization (equivalently, $-p(\mathbf{x})$ is convex). In geometry, the greatest and the least eigenvalues of $\mathbf{H}(\mathbf{x})$ measure the greatest and the least curvatures of $p(\mathbf{x})$, respectively [12]. In practice, the absolute values of both the greatest and the least curvature are bounded, by which we conclude that $\mathbf{H}(\mathbf{x})$ is both upper and lower bounded in the eigenvalue sense. Assumption 6.2 states this fact.

We define the cooperative source-seeking problem in the following text.

Problem 6.1   Under Assumptions 6.1 and 6.2, in a k-dimensional workspace with a scalar-valued distribution $p(\mathbf{x})$, design an algorithm to drive the center of a group of robots to the source $\mathbf{x}_s = \text{argmax}(p(\mathbf{x}))$ and simultaneously drive all robots modeled by (6.1) to a desired formation. The available information for the control algorithm of the ith robot is the sampled value $p(\mathbf{x}_i(t))$ at time t.

Remark 6.2   Problem 6.1 gives the definition on the cooperative source-seeking problem investigated in this chapter. It has essential differences from gradient-free swarm optimization methods, for example, particle swarm optimization [13], due to the constraints on the robot dynamics. Intuitively, the particle can move to any nearby position for the next step in the particle swarm optimization, while only the positions satisfying (6.1) are allowed to move following the definition in Problem 6.1, which makes it challenging over the gradient-free swarm optimization problems.

6.3   Source Seeking with All-to-All Communications

In this section, we present the algorithm for source seeking with all-to-all communications based on the cooperative gradient estimation.

6.3.1   Cooperative Estimation of Gradients

Since the robots are equipped with sensors for concentration measurement (instead of gradient), each robot needs to make estimation of gradient and the follows the gradient direction to the source. In this section, we use a LS estimator for gradient estimation.

Generally, the measurement of $p(\mathbf{x})$ are different for robots located at different positions. Our goal is to estimate the gradient at $\mathbf{x}_c(t)$, which is the center of the formation, that is, $\mathbf{x}_c(t) = \frac{1}{n}\sum_{i=1}^{n}\mathbf{x}_i(t)$, based on the sampling of $p(\mathbf{x})$ acquired by each robot. We use the following linear parameterization model to adaptively approximate the distribution of $p(\mathbf{x})$:

$$\hat{\mathbf{y}}(t) = \begin{bmatrix} \mathbf{X}(t) & \mathbf{1} \end{bmatrix} \theta(t) \tag{6.2}$$

with

$$\mathbf{y}(t) = \begin{bmatrix} p(\mathbf{x}_1(t)) \\ p(\mathbf{x}_2(t)) \\ \dots \\ p(\mathbf{x}_n(t)) \end{bmatrix}, \tag{6.3}$$

where $\mathbf{x}_1(t), \mathbf{x}_2(t), \dots, \mathbf{x}_n(t)$ are all $k \times 1$ vectors with k denoting the space dimension, $\theta(t)$ is the estimation parameter, which is a $(k+1) \times 1$ vector, $\hat{p}_i(\mathbf{x}(t))$ is the estimation of $p(\mathbf{x}(t))$ by the ith robot at time t, and $\mathbf{1}$ is a $n \times 1$ vector with all entries equal to 1. The estimation error is defined to be the difference between the estimation $\hat{\mathbf{y}}(t)$ and the measurement $\mathbf{y}(t) = \left[p(\mathbf{x}_1(t)), p(\mathbf{x}_2(t)), \dots, p(\mathbf{x}_n(t)) \right]^T = \left[p_1(t), p_2(t), \dots, p_n(t) \right]^T$. The LS estimator minimizes the norm of the estimation error, that is, to make $\hat{\mathbf{y}}(t) \approx \mathbf{y}(t)$. Solving $\hat{\mathbf{y}}(t) \approx \mathbf{y}(t)$ yields

$$\begin{aligned} \theta(t) &= \begin{bmatrix} \mathbf{X}(t) & \mathbf{1} \end{bmatrix}^{+} \mathbf{y}(t) \\ &= \begin{bmatrix} \mathbf{X}^T(t)\mathbf{X}(t) & \mathbf{X}^T(t)\mathbf{1} \\ \mathbf{1}^T\mathbf{X}(t) & n \end{bmatrix}^{-1} \begin{bmatrix} \mathbf{X}^T(t) \\ \mathbf{1}^T \end{bmatrix} \mathbf{y}(t), \end{aligned} \tag{6.4}$$

where $\begin{bmatrix} \mathbf{X}(t) & \mathbf{1} \end{bmatrix}^{+}$ is the pseudo-inverse of $\begin{bmatrix} \mathbf{X}(t) & \mathbf{1} \end{bmatrix}$. Therefore, we get the following parameter estimation:

$$\begin{aligned} \theta(t) &= \begin{bmatrix} \mathbf{X}^T(t)\mathbf{X}(t) & \mathbf{X}^T(t)\mathbf{1} \\ \mathbf{1}^T\mathbf{X}(t) & n \end{bmatrix}^{-1} \begin{bmatrix} \mathbf{X}^T(t) \\ \mathbf{1}^T \end{bmatrix} \mathbf{y}(t) \\ \hat{\mathbf{g}}_c(t) &= \begin{bmatrix} \mathbf{I} & \mathbf{0} \end{bmatrix} \theta(t), \end{aligned} \tag{6.5}$$

where $\hat{\mathbf{g}}_c(t)$ is the gradient estimation at the formation center $\mathbf{x}_c(t)$ at time t, $\mathbf{I}$ is a $k \times k$ identity matrix $\mathbf{1}$ is a k-row vector with all entries equal to 1, and $\mathbf{0}$ is a k-row vector with all entries equal to 0. This equation gives us an optimal estimation of $\hat{\mathbf{g}}_c(t)$ in the sense of LSs.

6.3.2 Control Law Design

To solve the cooperative source-seeking problem, we need to design two behaviors for the robot: one is the gradient climbing behavior, which drives the robot to the source; the other one is the formation achieving behavior, which guides robots to the desired formation. However, the goal is not realizable with only the two behaviors. In addition, we introduce a velocity damping behavior to avoid oscillation or overshooting around the source and an estimation error compensation behavior to reduce the effect of the gradient estimation error. Therefore, we present the following control input to the ith robot,

$$\mathbf{u}_i = - \sum_{j \in \mathbb{N}(i)} \omega_{1ij}(\mathbf{x}_i - \mathbf{x}_j - \mathbf{x}_{di} + \mathbf{x}_{dj}) + c_0 \hat{\mathbf{g}}_c$$

$$- \sum_{j \in \mathbb{N}(i)} \omega_{2ij}(\mathbf{v}_i - \mathbf{v}_j) - \frac{c_1}{n} \sum_{i=1}^{n} \mathbf{v}_i - c_2 \text{sgn} \left(\sum_{i=1}^{n} \mathbf{v}_i \right), \qquad (6.6)$$

where $\mathbb{N}(i)$ denotes the neighbor set of the ith robot; n denotes the number of robots in the group, $\omega_{1ij} = \omega_{1ji}$ and $\omega_{2ij} = \omega_{2ji}$, which are positive constants; c_0, c_1, and c_2 are also positive constants; $\text{sgn}(\cdot)$ is the sign function, which equals to 1, -1, and 0 for a positive input, negative input, and the input of 0, respectively; $\mathbf{x}_i$ and $\mathbf{v}_i$ are the ith robot's position and velocity, respectively; $\mathbf{x}_{di}$ is the desired relative position of the ith robot in the desired formation; and $\hat{\mathbf{g}}_c$ is the gradient estimation given by (6.5). In (6.6), the first two terms drive the robot to the desired formation; the third term generates the gradient climbing movement; the fourth term is a velocity damping term, which dissipates the kinematic energy of the robot; and the last term is an extra damping term to compensate the inaccuracy of gradient estimation. Combining the control input of all robots in the group, the control algorithm can be written into a compact form as follows:

$$\mathbf{u} = -(\mathbf{L}_1 \otimes \mathbf{I})(\mathbf{x} - \mathbf{x}_d) - (\mathbf{L}_2 \otimes \mathbf{I})\mathbf{v} + c_0 \mathbf{1} \otimes \hat{\mathbf{g}}_c$$

$$- \frac{c_1}{n} \mathbf{1} \otimes ((\mathbf{1}^T \otimes \mathbf{I})\mathbf{v}) - \frac{c_2}{n} \mathbf{1} \otimes \text{sgn}((\mathbf{1}^T \otimes \mathbf{I})\mathbf{v}). \qquad (6.7)$$

Here, $\mathbf{u} = [\mathbf{u}_1^T, \mathbf{u}_2^T, ..., \mathbf{u}_n^T]^T$ is the control input; $\mathbf{x} = [\mathbf{x}_1^T, \mathbf{x}_2^T, ..., \mathbf{x}_n^T]^T$ is the position vector of all robots; $\mathbf{v} = [\mathbf{v}_1^T, \mathbf{v}_2^T, ..., \mathbf{v}_n^T]^T$ is the velocity vector of all robots; both $\mathbf{L}_1$ and $\mathbf{L}_2$ are symmetric Laplacian matrices on the undirected graph constructed by the group of robots; the i–jth entry of $\mathbf{L}_1$ is $-\omega_{1ij}$ for $i \neq j$ and $\sum_{l=1}^{n} \omega_{1il}$ for $i = j$; the i–jth entry of $\mathbf{L}_2$ is $-\omega_{2ij}$ for $i \neq j$ and $\sum_{l=1}^{n} \omega_{2il}$ for $i = j$; $\mathbf{I}$ is an $k \times k$ identity matrix with k denoting the dimension of the space; n is the number of robots; $\mathbf{x}_d$ is a constant vector with $n \times k$ rows, which represents the desired formation; $\mathbf{1}$ is a $n \times 1$ vector with all entries equal to 1; and $\otimes$ is the Kronecker product.

The procedures of the proposed control algorithm with all-to-all communications is stated in Algorithm 6.1 for clarity.

In Algorithm 6.1, the ith robot first collects its measurement of concentration p_i, position $\mathbf{x}_i$, and velocity $\mathbf{v}_i$ (Line 2), and collects concentration, position, and velocity of all the other robots by communication (Line 3). After this, the position matrix $\mathbf{X}$ is constructed according to Equation (6.3) (Line 4). Then, gradient estimation of $\hat{\mathbf{g}}_c(t)$ is made according to Equation (6.5) (Line 5). Subsequently, Equation (6.6) is used to calculate the control input $\mathbf{u}_i$. Line 2, 3, 4, 5, and 6 are repeated in sequence until the norm of the estimated gradient $\hat{\mathbf{g}}_c(t)$ is less than a predefined positive constant ϵ, that is, $\|\hat{\mathbf{g}}_c(t)\| < \epsilon$.

Algorithm 6.1 Source-seeking control for the ith robot with all-to-all communication

---

Require:
  Concentration measurement p, position $\mathbf{x}$, velocity $\mathbf{v}$, desired formation $\mathbf{x}_d$, of all robots are available to the ith robot.
Ensure:
  To achieve the desired formation and drive the formation center to the source.
  1: **repeat**
  2:   $p_i, \mathbf{x}_i, \mathbf{v}_i \Leftarrow$ Sensor readings.
  3:   $p_1, p_2, ..., p_{i-1}, p_{i+1}, ..., p_n; \mathbf{x}_1, \mathbf{x}_2, ..., \mathbf{x}_{i-1}, \mathbf{x}_{i+1}, ..., \mathbf{x}_n$
       $\mathbf{v}_1, \mathbf{v}_2, ..., \mathbf{v}_{i-1}, \mathbf{v}_{i+1}, ..., \mathbf{v}_n$
       $\Leftarrow$ Communication with all the other robots.
  4:   Calculate position matrix $\mathbf{X}$.
  5:   Calculate gradient estimation $\hat{\mathbf{g}}_c(t)$.
  6:   Calculate control input $\mathbf{u}_i$.
  7: **until** $(\|\hat{\mathbf{g}}_c(t)\| < \epsilon)$

To validate the effectiveness of Algorithm 6.1 in theory, we have Assumption 6.3.

Assumption 6.3 The gradient estimation $\hat{\mathbf{g}}_c(t)$ obtained by (6.5) has a bounded error, that is, $\|\hat{\mathbf{g}}_c(t) - \mathbf{g}_c(t)\| \leq e_0$ (e_0 is a positive constant), in which $\mathbf{g}_c(t)$ denotes the true value of the gradient at the formation center.

We have Theorem 6.1 to state the convergence of the designed control algorithm.

Theorem 6.1 Under Assumptions 6.1, 6.2, 6.3, Algorithm 6.1 with the control law (6.6), where the parameter $c_2 > c_0 \sqrt{k} e_0$ (k denotes the dimension of space), solves Problem 6.1. The formation center $\mathbf{x}_c(t) = \frac{1}{n} \sum_{i=1}^{n} \mathbf{x}_i(t)$ converges to $\mathbf{x}^*$, which satisfies $\hat{\mathbf{g}}_c(\mathbf{x}^*) = \mathbf{0}$. Moreover, $\mathbf{x}^*$ has a bounded distance from the source $\mathbf{x}_s$:

$$\|\mathbf{x}^* - \mathbf{x}_s\| \leq \frac{2e_0}{\xi_1}, \tag{6.8}$$

where ξ_1 and e_0 are defined in Assumptions 6.2 and 6.3, respectively.

Proof: There are three steps for the proof: Step 1—derivation of the formation center's dynamics; Step 2—stability analysis using Lyapunov theory and Step 3—derivation of the source-seeking error.
  Step 1: derivation of the formation center's dynamics.
  Substituting the control input (6.7) into the robot dynamics (6.1) yields

$$\dot{\mathbf{v}} = -(\mathbf{L}_1 \otimes \mathbf{I})(\mathbf{x} - \mathbf{x}_d) - (\mathbf{L}_2 \otimes \mathbf{I})\mathbf{v} + c_0 \mathbf{1} \otimes \hat{\mathbf{g}}_c$$

$$- \frac{c_1}{n} \mathbf{1} \otimes \left((\mathbf{1}^T \otimes \mathbf{I})\mathbf{v} \right) - c_2 \mathbf{1} \otimes \text{sgn} \left((\mathbf{1}^T \otimes \mathbf{I})\mathbf{v} \right). \tag{6.9}$$

As to the formation center $\mathbf{x}_c$, we have

$$\mathbf{x}_c = \frac{1}{n} \sum_{i=1}^{n} \mathbf{x}_i = \frac{1}{n}(\mathbf{1}^T \otimes \mathbf{I})\mathbf{x}. \tag{6.10}$$

Recalling the property of Laplacian matrices that $\mathbf{L}_1 \mathbf{1} = \mathbf{0}$, the fact that $\mathbf{L}_1$ and $\mathbf{L}_2$ are symmetric, and the mixed-product property of Kronecker products, we have $(\mathbf{1}^T \otimes \mathbf{I})(\mathbf{L}_1 \otimes \mathbf{I}) = \mathbf{0}$, $(\mathbf{1}^T \otimes \mathbf{I})(\mathbf{L}_2 \otimes \mathbf{I}) = \mathbf{0}$, $(\mathbf{1}^T \otimes \mathbf{I})(\mathbf{1} \otimes \hat{\mathbf{g}}_c) = n\hat{\mathbf{g}}_c$, $(\mathbf{1}^T \otimes \mathbf{I})\left(\mathbf{1} \otimes \left((\mathbf{1}^T \otimes \mathbf{I})\dot{\mathbf{x}}\right)\right) = n^2 \dot{\mathbf{x}}_c$, $(\mathbf{1}^T \otimes \mathbf{I})\left(\mathbf{1} \otimes \text{sgn}\left((\mathbf{1}^T \otimes \mathbf{I})\dot{\mathbf{x}}\right)\right) = n\text{sgn}(\dot{\mathbf{x}}_c)$. Left multiplying the matrix $\frac{1}{n}(\mathbf{1}^T \otimes \mathbf{I})$ on both sides of (6.9), it yields the following dynamics of $\mathbf{x}_c$:

$$\ddot{\mathbf{x}}_c = c_0 \hat{\mathbf{g}}_c - c_1 \dot{\mathbf{x}}_c - c_2 \text{sgn}(\dot{\mathbf{x}}_c). \tag{6.11}$$

Defining $\mathbf{v}_c = \dot{\mathbf{x}}_c$, we have

$$\dot{\mathbf{v}}_c = c_0 \hat{\mathbf{g}}_c - c_1 \mathbf{v}_c - c_2 \text{sgn}(\mathbf{v}_c). \tag{6.12}$$

Step 2: stability analysis using the Lyapunov stability theory.

We choose the following Lyapunov function:

$$V = (\mathbf{x} - \mathbf{x}_d - \mathbf{1} \otimes \mathbf{x}_c)^T (\mathbf{L}_1 \otimes \mathbf{I})(\mathbf{x} - \mathbf{x}_d - \mathbf{1} \otimes \mathbf{x}_c) + p(\mathbf{x}_s) - p(\mathbf{x}_c)$$

$$+ \frac{1}{2c_0} \mathbf{v}_c^T \mathbf{v}_c + (\mathbf{v} - \mathbf{1} \otimes \mathbf{v}_c)^T (\mathbf{v} - \mathbf{1} \otimes \mathbf{v}_c).$$

Note that, in the expression of V, $\mathbf{L}_1 \otimes \mathbf{I}$ is symmetric since $(\mathbf{L}_1 \otimes \mathbf{I})^T = \mathbf{L}_1^T \otimes \mathbf{I}^T = \mathbf{L}_1 \otimes \mathbf{I}$, and $(\mathbf{x} - \mathbf{x}_d - \mathbf{1} \otimes \mathbf{x}_c)^T (\mathbf{L}_1 \otimes \mathbf{I})(\mathbf{x} - \mathbf{x}_d - \mathbf{1} \otimes \mathbf{x}_c)$ is semi-positive definite since the Laplacian matrix $\mathbf{L}_1$ is semi-positive definite [14] and the eigenvalues of $\mathbf{A} \otimes \mathbf{B}$ is $\lambda_i \mu_j$ for all i and j with λ_i denoting the ith eigenvalue of the square matrix $\mathbf{A}$ while μ_j denoting the jth eigenvalue of the square matrix $\mathbf{B}$. In addition, $p(\mathbf{x}_s) - p(\mathbf{x}_c) \geq 0$ due to Assumption 6.1. Therefore, V is indeed semi-positive definite. Calculating the time derivative of V along the trajectory of (6.9) and (6.12) yields

$$\dot{V} = 2(\mathbf{x} - \mathbf{x}_d)^T (\mathbf{L}_1 \otimes \mathbf{I})\mathbf{v} - \mathbf{g}_c^T \mathbf{v}_c + \frac{1}{c_0} \dot{\mathbf{v}}_c^T \mathbf{v}_c$$

$$+ 2(\dot{\mathbf{v}} - \mathbf{1} \otimes \dot{\mathbf{v}}_c)^T (\mathbf{v} - \mathbf{1} \otimes \mathbf{v}_c). \tag{6.13}$$

From (6.9) and (6.12), we get

$$\dot{\mathbf{v}} - \mathbf{1} \otimes \dot{\mathbf{v}}_c = -(\mathbf{L}_1 \otimes \mathbf{I})(\mathbf{x} - \mathbf{x}_d) - (\mathbf{L}_2 \otimes \mathbf{I})\mathbf{v}$$

$$- \frac{c_1}{n} \mathbf{1} \otimes \left((\mathbf{1}^T \otimes \mathbf{I})\mathbf{v} \right) - \frac{c_2}{n} \mathbf{1} \otimes \text{sgn}\left((\mathbf{1}^T \otimes \mathbf{I})\mathbf{v} \right)$$

$$+ c_1 \mathbf{1} \otimes \mathbf{v}_c + c_2 \mathbf{1} \otimes \text{sgn}(\mathbf{v}_c). \tag{6.14}$$

Along with (6.10), we have

$$\dot{\mathbf{v}} - \mathbf{1} \otimes \dot{\mathbf{v}}_c = -(\mathbf{L}_1 \otimes \mathbf{I})(\mathbf{x} - \mathbf{x}_d) - (\mathbf{L}_2 \otimes \mathbf{I})\mathbf{v} \tag{6.15}$$

Substituting (6.11) and (6.15) into (6.13) yields $\dot{V} = -2\mathbf{v}^T(\mathbf{L}_2 \otimes \mathbf{I})\mathbf{v} + (\hat{\mathbf{g}}_c - \mathbf{g}_c)^T \mathbf{v}_c - \frac{c_1}{c_0}\mathbf{v}_c^T\mathbf{v}_c - \frac{c_2}{c_0}\|\mathbf{v}_c\|_1$, where $\|\cdot\|_1$ denotes the 1-norm of a vector. Using properties of vector norms, we further get

$$\dot{V} \leq -2\mathbf{v}^T(\mathbf{L}_2 \otimes \mathbf{I})\mathbf{v} + \|\hat{\mathbf{g}}_c - \mathbf{g}_c\|_1 \|\mathbf{v}_c\|_1 - \frac{c_1}{c_0}\mathbf{v}_c^T\mathbf{v}_c - \frac{c_2}{c_0}\|\mathbf{v}_c\|_1$$

$$\leq -2\mathbf{v}^T(\mathbf{L}_2 \otimes \mathbf{I})\mathbf{v} - \frac{c_1}{c_0}\mathbf{v}_c^T\mathbf{v}_c - \left(\frac{c_2}{c_0} - \sqrt{k}e_0\right)\|\mathbf{v}_c\|_1$$

$$\leq 0. \tag{6.16}$$

Note that $\|\cdot\|$ denotes the Euclidean norm in (6.16). In the derivation of (6.16), the norm inequalities $(\hat{\mathbf{g}}_c - \mathbf{g}_c)^T \mathbf{v}_c \leq \|\hat{\mathbf{g}}_c - \mathbf{g}_c\|_1 \|\mathbf{v}_c\|_1$ and $\|\mathbf{v}_c\|_1 \leq \sqrt{k}\|\mathbf{v}_c\|$ are employed. The right side of (6.16) is semi-negative definite by noting that $\mathbf{L}_2 \otimes \mathbf{I}$ is semi-positive definite and $c_2 > c_0\sqrt{k}e_0$. Together with the fact that the right side of (6.16) is a function of state variables and the invariance-like theorem (Theorem 8.4 in Ref. [15]), we draw the conclusion that the right side of (6.16) goes to zeros as time elapses. Therefore,

$$\lim_{t\to\infty} \mathbf{v} = \mathbf{1} \otimes \boldsymbol{\alpha}_1 \text{ and } \lim_{t\to\infty} \mathbf{v}_c = \mathbf{0}, \tag{6.17}$$

where $\boldsymbol{\alpha}_1$ is a vector with k rows. With (6.11) and (6.15), we further conclude the following holds as time goes to infinity,

$$\hat{\mathbf{g}}_c \to \mathbf{0} \tag{6.18}$$

$$-(\mathbf{L}_1 \otimes \mathbf{I})(\mathbf{x} - \mathbf{x}_d) \to \mathbf{1} \otimes \dot{\boldsymbol{\alpha}}_1. \tag{6.19}$$

For any $k \times 1$ vector $\boldsymbol{\alpha}_2$, we have $\dot{\boldsymbol{\alpha}}_1^T \boldsymbol{\alpha}_2 = 0$ by left multiplication of $\mathbf{1}^T \otimes \boldsymbol{\alpha}_2^T$ on both sides of (6.19). Due to the arbitrariness of the choice of $\boldsymbol{\alpha}_2$, we get $\dot{\boldsymbol{\alpha}}_1 = \mathbf{0}$. In other words, we conclude that $\boldsymbol{\alpha}_1$ is a constant vector. With (6.19), we get $\mathbf{x} - \mathbf{x}_d \to \mathbf{1} \otimes \boldsymbol{\alpha}_3$ as $t \to \infty$, where $\boldsymbol{\alpha}_3$ is a vector with k rows. This equation means that, when time elapses, the position $\mathbf{x}$ is a translation from $\mathbf{x}_d$ given by the desired formation. Since transformation does not change relative positions, we obtain the conclusion that $\mathbf{x}$ converges to the formation given by $\mathbf{x}_d$. Moreover, we get $\mathbf{x}_c \to \mathbf{x}^*$ by employing $\hat{\mathbf{g}}_c(\mathbf{x}^*) = \mathbf{0}$ and (6.18).

Step 3: derivation of the source-seeking error.

The Taylor expansion of $p(\mathbf{x})$ at $\mathbf{x}^*$ yields $p(\mathbf{x}) = p(\mathbf{x}^*) + \nabla^T p(\mathbf{x}^*)(\mathbf{x} - \mathbf{x}^*) + \frac{1}{2}(\mathbf{x} - \mathbf{x}^*)^T H(\mathbf{x}_1)(\mathbf{x} - \mathbf{x}^*)$, where $\mathbf{x}_1$ is between $\mathbf{x}$ and $\mathbf{x}^*$. For $\mathbf{x} = \mathbf{x}_s$, this equation yields $p(\mathbf{x}_s) = p(\mathbf{x}^*) + \nabla^T p(\mathbf{x}^*)(\mathbf{x}_s - \mathbf{x}^*) + \frac{1}{2}(\mathbf{x}_s - \mathbf{x}^*)^T H(\mathbf{x}_1)(\mathbf{x}_s - \mathbf{x}^*)$. Thus, we get

$$0 \leq p(\mathbf{x}_s) - p(\mathbf{x}^*) \leq \|\nabla p(\mathbf{x}^*)\| \|\mathbf{x}_s - \mathbf{x}^*\| - \frac{\xi_1}{2}\|\mathbf{x}_s - \mathbf{x}^*\|^2$$

$$= \|\nabla p(\mathbf{x}^*) - \hat{\mathbf{g}}_c(\mathbf{x}^*)\| \|\mathbf{x}_s - \mathbf{x}^*\| - \frac{\xi_1}{2}\|\mathbf{x}_s - \mathbf{x}^*\|^2$$

$$\leq e_0 \|\mathbf{x}_s - \mathbf{x}^*\| - \frac{\xi_1}{2}\|\mathbf{x}_s - \mathbf{x}^*\|^2.$$

Note that Assumption 6.3 is employed in the given derivation of (6.20). Also note that the left side of (6.20) is not larger than 0 according to Assumption 6.1, so we have $0 \leq e_0 \|\mathbf{x}_s - \mathbf{x}^*\| - \frac{\xi_1}{2}\|\mathbf{x}_s - \mathbf{x}^*\|^2$. Therefore, $\|\mathbf{x}^* - \mathbf{x}_s\| \leq \frac{2e_0}{\xi_1}$. This concludes the proof. □

6.4 Distributed Source Seeking with Limited Communications

In Section 6.3, we developed a control algorithm for cooperative source seeking. It requires all-to-all communications. Aiming at reducing communication burdens, in this section we develop a fully distributed control algorithm, which only requires neighbor-to-neighbor communications. That is, we impose Assumption 6.4 in this section:

Assumption 6.4 The communication topology is a connected undirected graph. For the ith robot, only information from its one-hop neighbors and that from itself are available to the control design of $\mathbf{u}_i$.

From (6.6), it is clear that all necessary information for the ith robot can be derived from its one-hop neighbors except $\hat{\mathbf{g}}_c$ and $\frac{1}{n}(\mathbf{1}^T \otimes \mathbf{I})\mathbf{v}$. We use consensus filters to estimate them in a distributed manner [14, 16]. With consensus filters, a robot is able to estimate the average of inputs, that is, $\frac{1}{n}\sum_j^n \tau_j$, by running the following protocol on every robot

$$\dot{z}_i = \sum_{j \in \mathbb{N}(i)} a_{ij}(z_j - z_i) + \gamma(\tau_i - z_i), \tag{6.20}$$

where z_i is a scalar state maintained by the ith robot; $\mathbb{N}(i)$ denotes the neighbor set of the ith robot; and a_{ij} is a positive constant, which satisfies $a_{ij} = a_{ji}$. γ is a positive constant and u_i is the scalar input to the ith robot. By running (6.20) on every robot, z_i is able to track the average of inputs, that is, $\frac{1}{n}\sum_j^n \tau_j$. To estimate $\hat{\mathbf{g}}_c$ in distributed manners, we first rewrite the expression of $\theta(t)$ in (6.5) into average forms (without confusions, the time t aside the time-varying variables are omitted).

$$
\theta = \begin{bmatrix} \frac{1}{n}\mathbf{X}^T\mathbf{X} & \frac{1}{n}\mathbf{X}^T\mathbf{1} \\ \frac{1}{n}\mathbf{1}^T\mathbf{X} & 1 \end{bmatrix}^{-1} \begin{bmatrix} \frac{1}{n}\mathbf{X}^T\mathbf{y} \\ \frac{1}{n}\mathbf{1}^T\mathbf{y} \end{bmatrix}
$$

$$
= \begin{bmatrix} \frac{1}{n}\sum_{i=1}^n \mathbf{x}_i\mathbf{x}_i^T & \frac{1}{n}\sum_{i=1}^n \mathbf{x}_i \\ \frac{1}{n}\sum_{i=1}^n \mathbf{x}_i^T & 1 \end{bmatrix}^{-1} \begin{bmatrix} \frac{1}{n}\sum_{i=1}^n \mathbf{x}_i p_i \\ \frac{1}{n}\sum_{i=1}^n p_i \end{bmatrix}. \tag{6.21}
$$

We can online estimate $\frac{1}{n}\sum_{i=1}^n \mathbf{x}_i\mathbf{x}_i^T$, $\frac{1}{n}\sum_{i=1}^n \mathbf{x}_i$, $\frac{1}{n}\sum_{i=1}^n \mathbf{x}_i p_i$, and $\frac{1}{n}\sum_{i=1}^n p_i$ distributively by running four separate consensus filters on every robot.

We have the following filter expressions for $\mathbf{Z}_{1i}$, z_{2i}, z_{3i}, z_{4i}, θ_{ei}, and $\hat{\mathbf{g}}_{cei}$, which are estimations of $\frac{1}{n}\sum_{i=1}^n \mathbf{x}_i\mathbf{x}_i^T$, $\frac{1}{n}\sum_{i=1}^n \mathbf{x}_i$, $\frac{1}{n}\sum_{i=1}^n \mathbf{x}_i p_i$, $\frac{1}{n}\sum_{i=1}^n p_i$, θ, and $\hat{\mathbf{g}}_c$ by the *i*th robot (note that $\mathbf{Z}_{1i}$ is a $k \times k$ matrix, $\mathbf{z}_{2i}$ and $\mathbf{z}_{3i}$ are both $k \times 1$ vectors, and z_{4i} is a scalar):

$$\dot{\mathbf{Z}}_{1i} = \sum\nolimits_{j \in \mathbb{N}(i)} a_{ij}(\mathbf{Z}_{1j} - \mathbf{Z}_{1i}) + \gamma(\mathbf{x}_i \mathbf{x}_i^T - \mathbf{Z}_{1i})$$

$$\dot{\mathbf{z}}_{2i} = \sum\nolimits_{j \in \mathbb{N}(i)} a_{ij}(\mathbf{z}_{2j} - \mathbf{z}_{2i}) + \gamma(\mathbf{x}_i - \mathbf{z}_{2i})$$

$$\dot{\mathbf{z}}_{3i} = \sum\nolimits_{j \in \mathbb{N}(i)} a_{ij}(\mathbf{z}_{3j} - \mathbf{z}_{3i}) + \gamma(\mathbf{x}_i p_i - \mathbf{z}_{3i})$$

$$\dot{z}_{4i} = \sum\nolimits_{j \in \mathbb{N}(i)} a_{ij}(z_{4j} - z_{4i}) + \gamma(p_i - z_{4i})$$

$$\theta_{ei} = \begin{bmatrix} \mathbf{Z}_{1i} & \mathbf{z}_{2i} \\ \mathbf{z}_{2i}^T & 1 \end{bmatrix}^{-1} \begin{bmatrix} \mathbf{z}_{3i} \\ z_{4i} \end{bmatrix}$$

$$\hat{\mathbf{g}}_{cei} = \begin{bmatrix} \mathbf{I} & \mathbf{0} \end{bmatrix} \theta_{ei}. \tag{6.22}$$

To estimate $\frac{1}{n}(\mathbf{1}^T \otimes \mathbf{I})\mathbf{v}$, we first express it into the form that $\frac{1}{n}(\mathbf{1}^T \otimes \mathbf{I})\mathbf{v} = \frac{1}{n}\sum_{i=1}^{n} \mathbf{v}_i$. Denoting $\mathbf{z}_{5i}$ the estimation of $\frac{1}{n}(\mathbf{1}^T \otimes \mathbf{I})\mathbf{v}$ by the *i*th robot, we have

$$\dot{\mathbf{z}}_{5i} = \sum_{j \in \mathbb{N}(i)} a_{ij}(\mathbf{z}_{5j} - \mathbf{z}_{5i}) + \gamma(\mathbf{v}_i - \mathbf{z}_{5i}). \tag{6.23}$$

Replacing $\hat{\mathbf{g}}_c$ and $\frac{1}{n}(\mathbf{1}^T \otimes \mathbf{I})\mathbf{v}$ in (6.6) with $\hat{\mathbf{g}}_{cei}$ in (6.22) and $\mathbf{z}_{5i}$ in (6.23), the distributed control algorithm for the *i*th robot writes

$$\mathbf{u}_i = -\sum_{j \in \mathbb{N}(i)} \omega_{1ij}(\mathbf{x}_i - \mathbf{x}_j - \mathbf{x}_{vi} + \mathbf{x}_{vj}) + c_0 \hat{\mathbf{g}}_{cei}$$

$$-\sum_{j \in \mathbb{N}(i)} \omega_{2ij}(\mathbf{v}_i - \mathbf{v}_j) - c_1 \mathbf{z}_{5i} - c_2 \mathrm{sgn}(\mathbf{z}_{5i}). \tag{6.24}$$

The procedures of the proposed control algorithm with limited communications is stated in Algorithm 6.2.

Algorithm 6.2 Distributed source-seeking control for the *i*th robot with limited communications

---

Require:
   Concentration measurement p, position $\mathbf{x}$, velocity $\mathbf{v}$, and the desired formation $\mathbf{x}_d$ of itself and its neighbors are available to the *i*th robot.
Ensure:
   To achieve the desired formation and drive the formation center to the source.
1: Initializing state variables $\mathbf{Z}_{1i}, \mathbf{z}_{2i}, \mathbf{z}_{3i}, z_{4i}, \mathbf{z}_{5i}$.
2: **repeat**
3:    $\mathbf{x}_i, \mathbf{v}_i, \mathbf{p}_i \Leftarrow$ Sensor readings.
4:    $\mathbf{x}_j, \mathbf{v}_j \Leftarrow$ Communication with neighbor $j \in \mathbb{N}(i)$.
5:    Collect *j*th robot's consensus filter output $\mathbf{Z}_{1j}, \mathbf{z}_{2j}, \mathbf{z}_{3j}, z_{4j}, \mathbf{z}_{5j}$.
6:    Calculate state update of consensus filter $\mathbf{Z}_{1i}, \mathbf{z}_{2i}, \mathbf{z}_{3i}, z_{4i}, \mathbf{z}_{5i}$
7:    Calculate distributed gradient estimation $\theta_{ei}, \hat{\mathbf{g}}_{cei}$
8:    Calculate distributed control input $\mathbf{u}_i$
9: **until** $(\|\hat{\mathbf{g}}_{cei}(t)\| < \epsilon)$

The difference of Algorithms 6.1 and 6.2 lies in that consensus filters are used to estimate $\hat{\mathbf{g}}_{cei}$ in a distributed way in Algorithm 6.2 (Line 5, 6, and 7).

Remark 6.3   In this section, we present a distributed control algorithm for cooperative source seeking. The core idea is to use distributed consensus filters running on a fast enough time scale to estimate the centralized quantities in Algorithm 6.1 and use the estimated values to replace the actual centralized quantity in the control.

Remark 6.4   For a n robot network with all-to-all communication, each robot communicates with all other $(n-1)$ ones, which results in a communication complexity of $O(n^2)$. With the increase of the number of robot n, the communication burden increases significantly. In contrast, each robot only communicates with a few neighboring robots in the limited communication case. For instance, for a communication graph with a two-dimensional (2D) lattice topology, each robot only communicates with four neighbors even when the total number of robots n increases; thus, the communication complexity is $O(n)$ in this system. Therefore, the proposed distributed control in limited communication is scalable.

6.5   Simulations

In this section, we compare our algorithms with the methods proposed in Ref. [10] and the method proposed in Ref. [8]. There are two methods proposed in Ref. [10]: one uses a single robot to perform the task and the other one uses a group of robots. We call the two strategies PGS and PGM for short, respectively, and we call the strategy proposed in Ref. [8] ES method for short.

Simulations are performed under a representative set of parameters. For the ES method, parameters are chosen to be the same as in Ref. [8]. For the PGS and PGM method, parameter setup cannot be found in the associated paper [10]. We choose $k_d = 5$, $k_s = 1$, $d_0 = 1$, $\kappa = 1$ (see that paper for definitions of each parameter), and the potential function is chosen to be the one suggested in the paper. For our methods, Algorithms 6.1 and 6.2, we choose $c_0 = 20$, $c_1 = 7$, $c_2 = 7$, $\gamma = 1$, $L_1 = 6L_0$, $L_2 = L_0$, $L_3 = 30L_0$ with

$$
L_0 = \begin{bmatrix} 2 & 0 & -1 & -1 & 0 \\ 0 & 2 & -1 & 0 & -1 \\ -1 & -1 & 3 & -1 & 0 \\ -1 & 0 & -1 & 2 & 0 \\ 0 & -1 & 0 & 0 & 1 \end{bmatrix}.
$$

For the proposed methods and the PGM method, five robots are employed. Robots start from different positions. For PGS and ES method, which use a single robot, the initial position is set to be the center of robots in the multiple robot case. For simplicity, we choose $p(\mathbf{x}) = 100 - \|\mathbf{x}\|$ [8, 10].

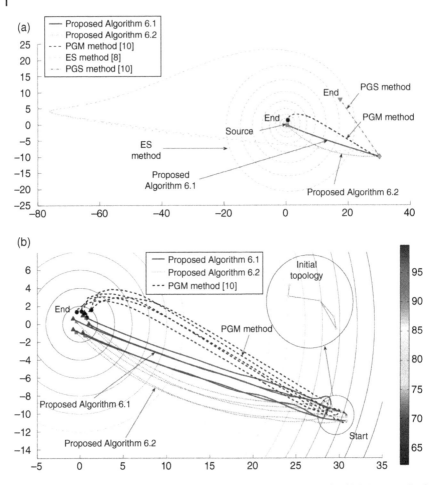

Figure 6.1 Simulation comparisons between PGS method [10], ES method [8], PGM method [10], and our Algorithms 6.1 and 6.2. The color bar denotes the concentration value of the source. (a) Trajectory of formation centers (for single robot case, it is the robot's own trajectory). The yellow diamond represents the start position. The red square is the end position of the robot by ES method and the proposed Algorithms 6.1 and 6.2. It is also the source position. The black dot and the pink triangle are the end position of the robot by PGM method and that of the robot by PGS method. (b) Trajectories of each robot for methods using multiple robots. The black dots, the blue triangles and the red squares are the end positions of robots by PGM method and Algorithms 6.1 and 6.2, respectively. The green dots represents the start positions. The contour of the field $p(\mathbf{x})$ is plotted in the figure. (*See insert for color representation of the figure.*)

We first compare the trajectories of the formation center (for methods using a single robot, we use its trajectory for evaluation). From Figure 6.1 (a), we can see that Algorithms 6.1 and 6.2 and PGM method and EM methods are able to steer the formation center to the vicinity of the source. Among them, ES method uses a single robot, which uses fewer robots in number than

Algorithms 6.1 and 6.2 and the PGM method. However, the robot takes a spiral like trajectory to the source and the traveled distance is much longer than other methods. Algorithm 6.1 outperforms Algorithm 6.2 and PGM and EM methods according to the traveled distance. Compared to Algorithm 6.1, the trajectory by using Algorithm 6.2 is a little longer, resulting from the dynamic interaction of the consensus filter and the robot dynamics in Algorithm 6.2. However, different from all the other methods simulated here, as stated before, Algorithm 6.2 is a fully distributed algorithm, which only requires information exchanges between one-hop neighbors. Nevertheless, Algorithm 6.2 still outperforms PGS, PGM, and ES methods in the sense that Algorithm 6.2 has a shorter trajectory than them.

We then compare Algorithms 6.1 and 6.2 and the PGM method, which use a group of robots, to see whether the desired formation are reached. From Figure 6.1(b), we can see that PGM method does not reach a uniform distribution on a circle, while Algorithms 6.1 and 6.2 reach the desired formation. The robot velocity profiles of our proposed methods are shown in Figures 6.2 and 6.3. As observed, the velocity of robots converges to a common value, which eventually reaches zero. This indicates that robots reach a common velocity in order to reach a fixed formation, while this common velocity converges to zero when the source is reached.

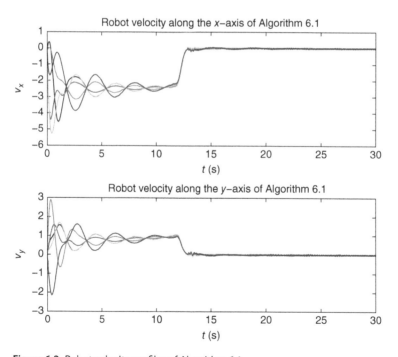

Figure 6.2 Robot velocity profiles of Algorithm 6.1.

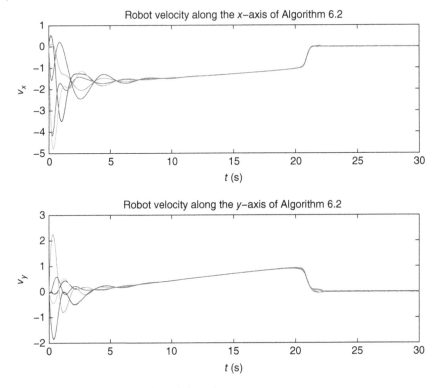

Figure 6.3 Robot velocity profiles of Algorithm 6.2.

6.6 Experimental Validation

In this section, the proposed algorithms are tested by robot experiments using E-puck robots in a light source-seeking application. The experimental implementation of the algorithms and the experimental results are described next.

6.6.1 The Robot

In the experiment, the E-puck mobile robot [17] is used to serve as the robot platform. It is a differential-driven wheeled mobile robot equipped with two step motors for movement, Bluetooth module for communication, and an onboard microprocessor for programming and control. Navigation sensors including infrared (IR) sensors, a VGA camera, and a 3D accelerometer are equipped on the robot. Encoders are equipped to record the relative pose (position and bearing). The kinematic model of the ith E-puck robot is

$$\begin{bmatrix} \dot{x}_i \\ \dot{y}_i \\ \dot{\alpha}_i \end{bmatrix} = \begin{bmatrix} \dfrac{\cos \alpha_i}{2} & \dfrac{\cos \alpha_i}{2} \\ \dfrac{\sin \alpha_i}{2} & \dfrac{\sin \alpha_i}{2} \\ -\dfrac{1}{l} & \dfrac{1}{l} \end{bmatrix} \begin{bmatrix} \Gamma_{i1} \\ \Gamma_{i2} \end{bmatrix}, \tag{6.25}$$

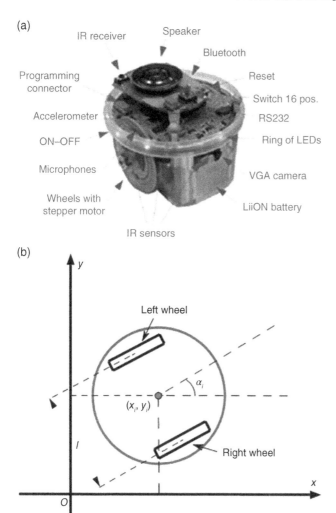

Figure 6.4 (a) The E-puck robot and (b) the system model.

where (x_i, y_i) represents the Cartesian coordinates of the middle point of the driving wheel axle; α_i is the bearing of the robot body with respect to the x-axis, l is the length between the two driving wheels; and Γ_{i1} and Γ_{i2} are the speeds of the left and the right wheels, respectively. Figure 6.4 shows the E-puck robot and its model.

Using feedback linearization, we are able to transform the robot model (6.25) into a new coordinates,

$$\begin{bmatrix} \dot{x}_i' \\ \dot{y}_i' \end{bmatrix} = \begin{bmatrix} \tau_{i1} \\ \tau_{i2} \end{bmatrix}, \tag{6.26}$$

where τ_{i1} and τ_{i2} are the new inputs, and

$$\begin{bmatrix} x_i' \\ y_i' \end{bmatrix} = \begin{bmatrix} x_i + c\cos\alpha_i \\ y_i + c\sin\alpha_i \end{bmatrix}, \tag{6.27}$$

with c a positive constant. The relationship between the wheel velocity input Γ_i and new input τ_i is expressed as

$$\begin{bmatrix} \Gamma_{i1} \\ \Gamma_{i2} \end{bmatrix} = M(\alpha_i)\begin{bmatrix} \tau_{i1} \\ \tau_{i2} \end{bmatrix}, \tag{6.28}$$

with

$$M(\alpha_i) = \begin{bmatrix} \frac{l}{2c}\sin\alpha_i + \cos\alpha_i & -\frac{l}{2c}\cos\alpha_i + \sin\alpha_i \\ -\frac{l}{2c}\sin\alpha_i + \cos\alpha_i & +\frac{l}{2c}\cos\alpha_i + \sin\alpha_i \end{bmatrix}. \tag{6.29}$$

For the E-puck robot, the length between wheels is $l = 0.053$ m. Also, we set $c = 0.0265$ m for our experimental validation.

6.6.2 The Experiment Setup

Light Source: A standard 75 W incandescent light bulb is placed on a big table locating at $(0.59, -1.36)$ in the global coordinates (the x-axis is pointing up and the y-axis is pointing left, with the origin at the center of the workspace).

Sensors: The E-puck robot has eight IR sensors, which are placed around the top of it, and are used to measure the light intensity. Closer to the light source, the sensor gets higher light intensity measurement values. To balance the light intensity measured by sensors from all directions, each E-puck robot utilizes the averaged value of its eight IR sensors to measure the local light intensity.

Localization System: The positions and bearings of the E-puck robots are obtained by odometry based on the information recorded by encoders with their initial positions and bearings calibrated in the global coordinate.

Robot Communication: The communication type among robots is set at "radio mode" (channel 1) with a communication range covering all peers. In the all-to-all communication case, the information from all other robots is used for state update, while the communication packages from the non-neighboring robots are intentionally dropped in the limited communication case to form the limited communication topology. The supervisory mode is implemented on E-pucks in the experiment.

Three robots, E-puck1, E-puck2, and E-puck3, are deployed in the light field. The initial relative positions of E-puck robots are randomly chosen as $(0, -2)$, $(-0.4, -1.7)$, and $(-0.2, -2)$ with bearings 0, 0, and 0, respectively, in the global coordinates. The virtual structure of the desired formation is set as $(x_{d1}, y_{d1}) = (0, 0.2)$, $(x_{d2}, y_{d2}) = (0.1, 0.1732)$, and $(x_{d3}, y_{d3}) = (0, 0)$, which corresponds to an equilateral triangle with edge length 0.2 m. The control coefficient c_0 is chosen as $c_0 = 10$. According to the communication topologies

(a)

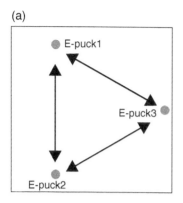

(b)

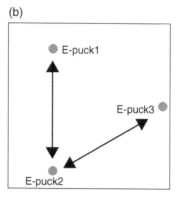

Figure 6.5 E-puck communication topologies considered in the experiment: (a) all-to-all communication and (b) limited communication.

shown in Figure 6.5, the Laplacian matrices in the all-to-all communication case and the limited communication case are, respectively, chosen as follows:

$$
\begin{bmatrix} 12 & -6 & -6 \\ -6 & 12 & -6 \\ -6 & -6 & 12 \end{bmatrix}, \quad \begin{bmatrix} 6 & -6 & 0 \\ -6 & 12 & -6 \\ 0 & -6 & 6 \end{bmatrix}
$$

6.6.3 Experimental Results

The experiment results are reported for both the all-to-all communication case and the limited communication case. The performance in these two scenarios are then compared.

6.6.3.1 All-to-All Communication

Algorithm 6.1 is implemented for the all-to-all communication case. The communication topology of three robots is shown in Figure 6.5(a). Figure 6.6 displays the experimental results. The initial positions of E-puck1, E-puck2, and E-puck3 are randomly chosen, and the robots move toward the light source. At time $t = 11$ s, the robots are still in the process of forming the equilateral triangle during their movement to the light source. The snapshot at time $t = 30$ s reveals that this process continues. At time $t = 51$ s, the equilateral triangle formation is almost reached. Then, the robots keep the formation, and gradually approach the light source, as shown in Figure 6.6 at $t = 65$ s. Finally, the formation center of the robots reaches the light source at $t = 84$ s as shown in Figure 6.6. Figure 6.7 shows the light intensity measurement of each robot increases as the robots approach to the light source, where noises and measurement uncertainties exist.

6.6.3.2 Limited Communication

The limited communication algorithm is tested under the same setup as in the case of all-to-all communication except the communication topology

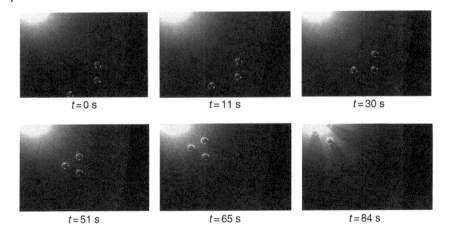

Figure 6.6 Light source-seeking experiment with the all-to-all communication on three E-puck robots.

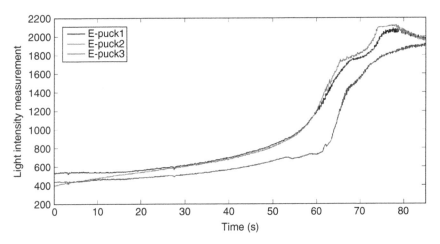

Figure 6.7 Robots' light intensity measurements in the all-to-all communication case. (*See insert for color representation of the figure.*)

is as shown in Figure 6.5(b). This implementation follows the procedure of Algorithm 6.2. Figure 6.8 shows the experimental results with limited communication. The three robots start from a dark place with a low light intensity value at $t = 0$ s. The robots move toward the desired formation and simultaneously move toward the light source at $t = 11$ s and $t = 30$ s. At time $t = 51$ s, the robots are still away from the desired formation. At time $t = 67$ s, the equilateral formation is almost formed. They maintain the formation until they finally reach the light source at $t = 91$ s.

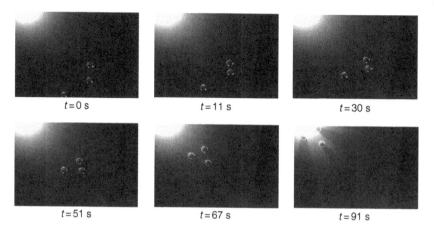

Figure 6.8 Light source-seeking experiment with the limited communication on three E-puck robots.

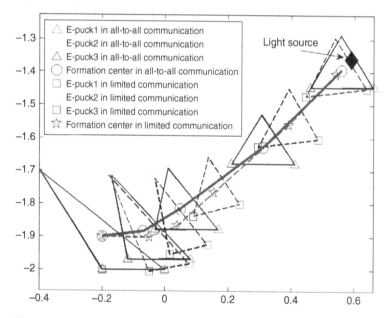

Figure 6.9 Trajectories of multirobot light source seeking on the E-puck platform with both the all-to-all and the limited communications. The robot positions and formation center are drawn in every 21 s (except that the last two dashed triangles are drawn in 7 s to show the end positions of the robots in limited communication).

6.6.3.3 Performance Comparison

Figure 6.9 shows odometry trajectories in both the all-to-all and the limited communication cases. The solid-line trajectory shows the robot center trajectory in all-to-all communication, and the dashed-line trajectory shows the robot center trajectory in limited communication. Comparing the all-to-all and the limited communication cases, it can be seen that the robot source seeking in the limited communication case takes longer time to converge to the source due to the use of consensus filter. Also, the path length in the limited communication case is slightly longer.

6.7 Conclusion

Cooperative distributed source seeking was studied in this chapter. Two control algorithms were proposed to solve the problem with all-to-all and limited communications, respectively. In the case of all-to-all communication, theoretical analysis proved that the proposed algorithm guarantees convergence to the source with all robots reaching the desired formation. With the constraint of limited communication, consensus filters were used to distributively estimate centralized quantities, and then embedded in the distributed control laws. MATLAB simulations in an ideal environmental setup demonstrated the effectiveness of the proposed algorithms. Experiments on the E-puck robot platform were performed to show the applicability of the proposed algorithms in a light source-seeking scenario.

Notes

The results in this chapter were first published in Ref. [18].

Acknowledgment is given to ©2014 IEEE. Reprinted, with permission, from Li, S., Kong, R., and Guo, Y. (2014) Cooperative distributed source seeking by multiple robots: Algorithms and experiments. IEEE/ASME Transactions on Mechatronics, 19 (6), 1810–1820.

References

1 Senga, H., Kato, N., Ito, A., Niou, H., Yoshie, M., Fujita, I., Igarashi, K., and Okuyama, E. (2007) Development of spilled oil tracking autonomous buoy system, Conference Paper. pp. 1–10, doi:10.1109/OCEANS.2007.4449148.
2 Li, W., Farrell, J., Pang, S., and Arrieta, R. (2006) Moth-inspired chemical plume tracing on an autonomous underwater vehicle. *IEEE Transactions on Robotics*, **22** (2), 292–307.
3 Sugawara, K., Kazama, T., and Watanabe, T. (2004) Foraging behavior of interacting robots with virtual pheromone, in *IEEE/RSJ International Conference on Intelligent Robots and Systems*, vol. 3, pp. 3074–3079.

4 Passino, K. (2002) Biomimicry of bacterial foraging for distributed optimization and control. *Control Systems Magazine, IEEE*, **22** (3), 52–67.

5 Farrell, J., Pang, S., and Li, W. (2005) Chemical plume tracing via an autonomous underwater vehicle. *IEEE Journal of Oceanic Engineering*, **30** (2), 428–442.

6 Ishida, H., Tanaka, H., Taniguchi, H., and Moriizumi, T. (2004) Mobile robot navigation using vision and olfaction to search for a gas/odor source, in *IEEE/RSJ International Conference on Intelligent Robots and Systems*, Sendai, Japan, vol. 1, pp. 313–318.

7 Hayes, A., Martinoli, A., and Goodman, R. (2002) Distributed odor source localization. *IEEE Sensors Journal*, **2** (3), 260–271.

8 Zhang, C., Siranosian, A., and Krstic, M. (2007) Extremum seeking for moderately unstable systems and for autonomous vehicle target tracking without position measurements. *Automatica*, **43** (10), 1832–1839.

9 Gazi, V. and Passino, K. (2004) Stability analysis of social foraging swarms. *IEEE Transactions on Systems, Man and Cybernetics*, **34**, 539–557.

10 Bachmayer, R. and Leonard, N. (2002) Vehicle networks for gradient descent in a sampled environment, in *IEEE Conference on Decision and Control*, vol. 1, pp. 112–117, doi:10.1109/CDC.2002.1184477.

11 Ogren, P., Fiorelli, E., and Leonard, N. (2004) Cooperative control of mobile sensor networks: Adaptive gradient climbing in a distributed environment. *IEEE Transactions on Automatic Control*, **49**, 1292–1302.

12 Lopez, R. (2008) Differential geometry of curves and surfaces in lorentz-minkowski space. *Notes of Mini-Course of University of Sao Paulo, Brasil, eprint arXiv:0810.3351*.

13 Kennedy, J. (2010) Particle swarm optimization, in *Encyclopedia of Machine Learning* (eds C. Sammut and G.I. Webb), Springer, Boston, MA, pp. 760–766.

14 Freeman, R., Yang, P., and Lynch, K. (2006) Stability and convergence properties of dynamic average consensus estimators, in *IEEE Conference on Decision and Control*, San Diego, CA, pp. 338–343.

15 Khalil, H. (2002) *Nonlinear Systems*, 3rd edn., Prentice Hall, Upper Saddle River, NJ.

16 Yang, P., Freeman, R., and Lynch, K. (2007) Distributed cooperative active sensing using consensus filters, in *IEEE International Conference on Robotics and Automation*, pp. 405–410, doi:10.1109/ROBOT.2007.363820.

17 Mondada, F., EPFL education robot: E-puck, http://www.e-puck.org/ (accessed April 2016).

18 Li, S., Kong, R., and Guo, Y. (2014) Cooperative distributed source seeking by multiple robots: Algorithms and experiments. *IEEE/ASME Transactions on Mechatronics*, **19** (6), 1810–1820.

7

Distributed Plume Front Tracking by Cooperative Robots

7.1 Introduction

The recent Deepwater Horizon oil spill has posed great challenges to both robotics and ocean engineering communities. It took months to estimate the extent of the underwater plume, and the accuracy of these estimates will likely be debated for years to come. The challenges motivate us to consider utilizing advanced robotic techniques to monitor and track the propagation of oil plumes. In this chapter, we propose a model-based method to track the dynamic plume front using multirobot platforms.

Existing approaches related to plume tracking can be categorized into the following three classes: mapping-based approach, behavior-based approach, and control-based approach. In the mapping-based approach, a concentration map of the environment is first built, and then plume tracking is conducted based on the obtained map. Representative work includes Ref. [1], where a grid map was created to represent the concentration distribution based on data collected by a single robot. In Ref. [2], a hidden Markov method (HMM) was employed to map the environment. Similar strategies were applied in Ref. [3] to localize a chemical plume source. In Ref. [4], the likelihood grid mapping method was extended to map an environment with multiple chemical sources. As it is necessary to scan the environment for mapping, it introduces extra overhead of robot scanning trajectory design, which often makes the map-based method time costly. On the contrary, the behavior-based approach combines certain elementary behaviors to track an emergent plume. Li et al. [5] developed bio-inspired chemical plume tracing on an autonomous underwater vehicle using elementary behaviors such as finding plume and tracking-in and tracking-out plume in a subsumption architecture. Social potential fields were used in Ref. [6] to coordinate group behaviors, where experiments demonstrated a robotic swarm was driven to find a spill and reach a perimeter formation. In Ref. [7], the Braitenberg model was employed to map the sensor reading of concentration to robot motor reaction in environments with odor plumes, and the effectiveness was validated by experimental results. Generally speaking, the behavior-based approach is often intuitive, and it usually takes

Distributed Cooperative Control: Emerging Applications, First Edition. Yi Guo.
© 2017 John Wiley & Sons, Inc. Published 2017 by John Wiley & Sons, Inc.
Companion website: www.wiley.com/go/Guo/DistributedCooperativeControl

extensive simulations or experiments to validate the method [5–7]. Due to the lack of rigorousness such as convergency analysis, performances are difficult to guarantee; thus, a working example may be difficult to extend to other scenarios with different parameters.

The control-based approach explicitly designs robot control laws with provable convergence. In Ref. [8], the plume-tracking problem was modeled by a finite automaton and the hybrid control theory was employed in theoretical analysis. In Ref. [9], the robotic manifold tracking of coherent structures in flows was investigated. In Ref. [10], a level curve–tracking problem was solved in two-dimensional (2D) space using multisensor platforms. A Kalman filter was developed to incorporate historical sensor reading into control laws that guide the robot movement. This approach was later extended to solve the same problem with control uncertainties in Ref. [11] and in 3D space in Ref. [12]. However, due to the lack of a fixed concentration reference for the level curve tracking, this type of method is subject to concentration drifting in the presence of noises in sensor readings.

In this chapter, we study dynamic plume-tracking control using multiple robots. The approach falls in the category of control-based method, and the plume-tracking problem is solved in an estimation and control framework. We first discuss the single-robot case, where the plume front dynamics is derived using the advection–diffusion equation governing the plume propagation. Then an observer is designed to estimate the dynamic movement of the plume front, and a feedback control law is constructed to track the plume front. We then extend the single robot case to a multirobot scenario, where an additional behavior of formation along the plume front is added with robots' control laws explicitly given for a multirobot team in a nearest-neighbor communication topology. The algorithms designed in both the single- and multirobot cases are tested in simulation, which show satisfactory performances.

The contribution of this chapter is twofold. First, the proposed method utilizes the propagation model (i.e., the advection–diffusion model) of point-source pollution in marine environments, and incorporate the model into the controller design. Second, the designed control is analytically constructed with provable convergence, and numerical computation of partial differential equations is avoided for rapid response and real-time control.

Comparing to existing work [10, 12], we consider *dynamic* plume where the plume front propagates along time, but the work in Refs. [10, 12] only considers *static* level curve tracking. Also, we exploit the transport of pollution source in water, and incorporate the advection–diffusion model into our control design, while the work in Refs. [10, 12] mainly replies on the gradient information (either implicitly or explicitly obtained) for plume tracking. The dynamic nature of our solution avoids the drifting problem that may occur during the static-level curve tracking due to uncertainties or sensor noises. Also, we provide an analytic solution that does not numerically solve the complex plume dynamics as done in Ref. [13] to save computational expenses.

Organization: The rest of this chapter is organized as follows. The plume-tracking problem is stated in Section 7.2. Section 7.3 investigates the single-robot plume front tracking problem. Then, the algorithm for the single-robot plume tracking is extended to solve multirobot simultaneous plume front tracking and formation in Section 7.4. Simulations are presented in Section 7.5. Section 7.6 concludes the chapter.

7.2 Problem Statement

In this section, we present the plume model, assumptions made on the robot and environments, and state the control problems studied in the chapter.

Chemicals introduced in the marine environment through point-source pollution generally propagate through two transport mechanisms: advection and diffusion. Advection is the transport of the chemical due to the motion of the flow, and diffusion is the motion from areas of higher concentration to areas of lower concentration. The propagation of chemicals is modeled by the following advection–diffusion equation:

$$\frac{\partial c(x,t)}{\partial t} + v^T(x,t)\nabla c(x,t) = k\nabla^2 c(x,t), \tag{7.1}$$

where $c \geq 0, c \in \mathbb{R}$ is the chemical concentration, t denotes time, v is the advection velocity, $k > 0, k \in \mathbb{R}$ is the diffusion coefficient in a homogenous space, $\nabla c(x,t) = \frac{\partial c(x,t)}{\partial x}$ is the spatial gradient of $c(x,t)$, $\nabla^2 c(x,t) = \frac{\partial^2 c(x,t)}{\partial x_1^2} + \frac{\partial^2 c(x,t)}{\partial x_2^2}$ is the divergence of $c(x,t)$ in 2D space, and $\frac{\partial c(x,t)}{\partial t}$ is the partial derivative of $c(x,t)$ along the time axis.

In this chapter, we consider plume front tracking using unmanned surface vehicles with the following kinematic model [14] in 2D space,

$$\dot{x}_r = \begin{bmatrix} \cos\theta_r & -\sin\theta_r \\ \sin\theta_r & \cos\theta_r \end{bmatrix} \tau, \tag{7.2}$$

where $x_r = [x_{r1}, x_{r2}]^T$ represents the Cartesian coordinates of the robot, $\tau = [\tau_1, \tau_2]^T$ are the surge and sway velocities, respectively, defined in the body-fixed frame, θ_r is the heading angle of the vehicle. Note that the determinant of

$$\begin{bmatrix} \cos\theta_r & -\sin\theta_r \\ \sin\theta_r & \cos\theta_r \end{bmatrix}$$

is always 1 whatever the value of θ, meaning that its inverse always exists. The inverse matrix can be analytically solved as $\begin{bmatrix} \cos\theta_r & \sin\theta_r \\ -\sin\theta_r & \cos\theta_r \end{bmatrix}$. Equation (7.2) can be converted into a single-integrator model by defining a new control input metric $u = [u_1, u_2]^T$ satisfying,

$$\tau = \begin{bmatrix} \cos\theta_r & \sin\theta_r \\ -\sin\theta_r & \cos\theta_r \end{bmatrix} u \tag{7.3}$$

with which, the robot dynamics is reduced to a single integrator model,

$$\dot{x}_r = u. \tag{7.4}$$

For the single-robot plume-tracking problem, we assume that it has sensors to obtain local information, including its position and heading, and the chemical concentration information at its current position. For the multi-robot plume-tracking case, we assume in addition that each robot is able to access information from its one-hop neighbors by communication. We made Assumptions 7.1 and 7.2.

Assumption 7.1   The robot's onboard sensors obtain its position x_r, heading θ_r, and the following information at its position: the chemical concentration c_r, the gradient of the concentration ∇c_r, the divergence of the concentration gradient $\nabla^2 c_r$, and the flow velocity v_r.

Remark 7.1   Note that there may not exist sensors to directly measure the gradient and the divergence of the concentration. However, the values can be estimated either by counting the historical data of a single sensor [15], or by considering the readings from multiple spatially distributed sensors on a single robot [16].

Assumption 7.2   For multirobot plume tracking, two robots are assigned as boundary robots and all the other robots are assigned as follower robots. As shown in Figure 7.1 as an example, the two boundary robots do not access information from other robots, and the follower robot, that is, the ith robot ($1 < i < n$ with n denoting the total number of robots), only communicates with

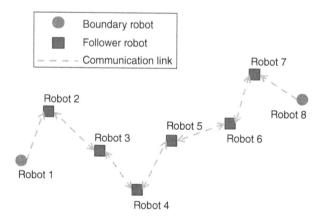

Figure 7.1 The assignment of boundary robots and follower robots in an eight-robot group, where the solid disk and the solid square represent the boundary robots and the follower robots, respectively; the dashed line represents the directional communication links between consecutive robots.

its one-hop neighbors on the communication graph, that is, the $(i-1)$th robot and the $(i+1)$th robot.

Remark 7.2   Assumption 7.2 significantly reduces the communication complexity by restricting the robot communication along a line topology with a communication complexity $O(n)$, in contrast to an all-to-all communication with a communication complexity $O(n^2)$ [17, 18].

At time t, we consider those points $x(t)$ with the concentration $c(x(t), t) = c_0$ (where c_0 is the threshold concentration) to be the plume front. We denote the plume front using the position set $\{x(t) \in \mathbb{R}^2, c(x(t), t) = c_0\}$. The basic control objective is to drive the robot to the plume front. To monitor the whole perimeter of the plume front, we impose the second objective for the single robot to patrol along the plume front with a desired speed, while in the multi-robot case, we want the robots distributed evenly along the plume front. The control objectives are formally stated in the following text.

Problem 7.1   **(Single-Robot Plume Front Tracking):** For the plume dynamics modeled by the advection–diffusion equation (7.1), design a control law under Assumption 7.1 to drive a single robot, which is subject to the dynamic constraint (7.4), to track the plume front $\{x(t) \in \mathbb{R}^2, c(x(t), t) = c_0\}$, and patrol along the plume front with a desired speed $v_d(t) \in \mathbb{R}$.

Problem 7.2   **(Distributed Multirobot Plume Front Tracking):** For the plume modeled by the advection–diffusion equation (7.1), design a control law under Assumptions 7.1 and 7.2 to drive a group of robots, which are subject to the dynamic constraint (7.4), to track the plume front $\{x(t) \in \mathbb{R}^2, c(x(t), t) = c_0\}$, and simultaneously reach an even distribution around the plume front.

7.3 Plume Front Estimation and Tracking by Single Robot

In this section, we solve Problem 7.1 for the single-robot plume tracking. The problem is modeled in an estimation and control perspective. A state equation and a measurement equation are constructed to design a nonlinear observer for the position estimation of a reference point on the plume front. Then, a control law is presented to steer the robot to the estimated position.

We partition the plume front tracking task into two parts: the first part is the plume front estimation part, which estimates the position of a reference point on the front; the second part is the tracking control part, which drives the robot to the estimated position. The estimation part and the tracking control part run simultaneously to reach the design objective. As illustrated in Figure 7.2,

Figure 7.2 The schematic illustration of the proposed algorithm. There are two steps in the algorithm: (i) the robot estimates the plume front trajectory and (ii) the robot tracks the estimated trajectory. The robot achieves front tracking if both the estimated and tracking trajectories converge.

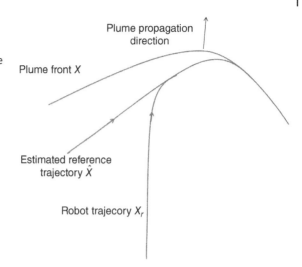

the estimation part maintains a variable $\hat{x}(t)$, which estimates the position of the reference point $x(t)$ on the plume front. The estimation part is designed to enable the convergence of $\hat{x}(t)$ to $x(t)$, that is, the convergence of the estimation error $e(t) = x(t) - \hat{x}(t)$ to zero as time elapses. The tracking control part steers the robot positioned at $x_r(t)$ to the estimated place $\hat{x}(t)$, to reduce the control error $e'(t) = \hat{x}(t) - x_r(t)$ along time. When both the estimation error $e(t)$ and the control error $e'(t)$ converge to zero, $x(t) - x_r(t) = e(t) + e'(t)$ will converge to zero and the design objective of driving the robot to the reference point is reached.

7.3.1 State Equation of the Plume Front Dynamics

In this section, we first derive the plume front dynamics based on the partial differential equation (7.1), and then use the plume front dynamics to construct the state equation of the reference point, which describes the expected behavior of the robot.

7.3.1.1 Plume Front Dynamics

In this section, we derive the plume front dynamics. Because the reference point describes the expected behavior of the robot, it locates on the plume front according to Assumption 7.1. Therefore, at time t, the reference point $x(t)$ satisfies

$$c(x(t), t) = c_0. \tag{7.5}$$

Without introducing confusion, we drop the variable t in $x(t)$ hereafter. Computing time derivative on both sides of Equation (7.5) yields

$$\dot{c}(x, t) = \frac{\partial c}{\partial t} + \nabla^T c \dot{x} = 0, \tag{7.6}$$

where $\frac{\partial c}{\partial t} = \frac{\partial c(x,t)}{\partial t}$ is the time derivative of the concentration and $\nabla c = \nabla c(x,t) = \frac{\partial c}{\partial x}$ is the spatial derivative, that is, the gradient of the concentration. With Equations (7.1) and (7.6), we get

$$\dot{x}^T \nabla c = -v_x^T \nabla c - k\nabla^2 c \tag{7.7}$$

where $v_x = v(x,t)$ and $\nabla^2 c = \nabla^2 c(x,t)$.

7.3.1.2 Derivation of the State Equation based on Plume Front Dynamics

Because the robot is also expected to patrol with speed $v_d = v_d(t)$ along the plume front, the reference point, which describes the ultimate behavior of the robot, is then subject to the following constraint:

$$\frac{(A\nabla c)^T}{\|\nabla c\|}\dot{x} = v_d, \tag{7.8}$$

where $A = \begin{bmatrix} 0 & -1 \\ 1 & 0 \end{bmatrix}$ is the orthogonal matrix and $A\nabla c$ represents a vector along the tangent direction of the plume front. Regarding Equations (7.7) and (7.8) as linear equations relative to $\dot{x}$, $\dot{x}$ can be uniquely solved as follows:

$$\dot{x} = -\frac{(v_x^T\nabla c + k\nabla^2 c)\nabla c}{\|\nabla c\|^2} + \frac{v_d A\nabla c}{\|\nabla c\|}. \tag{7.9}$$

7.3.2 Measurement Equation and Observer Design

For the robot positioned at x_r, the concentration field can be approximated by the first-order Taylor expansion as

$$c(x,t) = \nabla^T c_r(x - x_r) + c_r, \tag{7.10}$$

where x_r denotes the position of the robot at time t, $c_r = c(x_r, t)$ is the concentration at position x_r, and $\nabla c_r = \frac{\partial c(x,t)}{\partial x}\big|_{x=x_r}$ is the gradient at x_r. Define $y(x) = c(x,t)$ as the output of the measurement at time t. Then, the approximation (7.10) yields

$$y(x) = \nabla^T c_r(x - x_r) + c_r. \tag{7.11}$$

Note that the true value of $y(x)$ is $y(x) = c_0$ since the reference point x locates on the front. In addition, the measurement equals $y(\hat{x}) = c(\hat{x}) = \nabla^T c_r(\hat{x} - x_r)$ at the point $\hat{x}$.

Until now, we have obtained the state equation (7.9) and the measurement Equation (7.11), which rewrite as follows:

$$\dot{x} = -\frac{(v_x^T\nabla c + k\nabla^2 c)\nabla c}{\|\nabla c\|^2} + \frac{v_d A\nabla c}{\|\nabla c\|} \tag{7.12}$$

$$y(x) = \nabla^T c_r(x - x_r) + c_r. \tag{7.13}$$

For the system described by (7.12) and (7.13), we use the following extended Luenberger observer [19] for the state estimation:

$$\dot{\hat{x}} = -\frac{(v_{\hat{x}}^T \nabla c_{\hat{x}} + k\nabla^2 c_{\hat{x}})\nabla c_{\hat{x}}}{\|\nabla c_{\hat{x}}\|^2} + \frac{v_d A \nabla c_{\hat{x}}}{\|\nabla c_{\hat{x}}\|}$$
$$-k_3 \nabla c_r \left(\nabla^T c_r(\hat{x} - x_r) + c_r - c_0\right),$$
(7.14)

where $k_3 > 0$ is a coefficient, $v_{\hat{x}} = v_{\hat{x}}(\hat{x}, t)$, $\nabla c_{\hat{x}} = \nabla c(\hat{x})$, $\nabla^2 c_{\hat{x}} = \nabla^2 c(\hat{x})$.

Remarkably $v_{\hat{x}}$, $\nabla c_{\hat{x}}$, and $\nabla^2 c_{\hat{x}}$ are all quantities at position $\hat{x}$. For the robot at position x_r, they are neither measurable nor computable since these quantities, as functions of position $\hat{x}$ and time t, are unknown. For remedy, we simply replace them with the associated quantities at x_r, which are accessible by the robot according to Assumption 7.1. In this way, we have the following observer expression:

$$\dot{\hat{x}} = -\frac{(v_r^T \nabla c_r + k\nabla^2 c_r)\nabla c_r}{\|\nabla c_r\|^2} + \frac{v_d A \nabla c_r}{\|\nabla c_r\|}$$
$$-k_3 \nabla c_r \left(\nabla^T c_r(\hat{x} - x_r) + c_r - c_0\right),$$
(7.15)

where $v_r = v(x_r, t)$, $\nabla c_r = \nabla c(x_r, t)$ and $\nabla^2 c_r = \nabla^2 c(x_r, t)$.

Remark 7.3 Note that the replacement of $v_{\hat{x}}$, $\nabla c_{\hat{x}}$ and $\nabla^2 c_{\hat{x}}$ in (7.14) by v_r, ∇c_r, and $\nabla^2 c_r$ is reasonable provided the condition that $\hat{x} \to x_r$ with time. This condition is enforced by the active control law presented in Section 7.3.3.

Remark 7.4 The absolute value of $\left(\nabla^T c_r(\hat{x} - x_r) + c_r - c_0\right)\nabla c_r$ in Equation (7.15) goes small very fast when $\nabla^T c_r$ is small. To gain a fast enough response toward the desired reference point, it is necessary to tune k_3 large in this case. In contrast, when $\nabla^T c_r$ is large, the absolute value of this term becomes large very fast and thus needs a small k_3 as the gain. To adaptively adjust the gain k_3, we can set $k_3 = k_2/\|\left(\nabla^T c_r(\hat{x} - x_r) + c_r - c_0\right)\nabla c_r\|$ in implementation with $k_2 > 0$ as a constant to improve the performance.

7.3.3 Estimation-Based Tracking Control

For the robot model (7.4), we present the following control law to reduce the control error $e' = \hat{x} - x_r$:

$$u = -\frac{(v_r^T \nabla c_r + k\nabla^2 c_r)\nabla c_r}{\|\nabla c_r\|^2} + \frac{v_d A \nabla c_r}{\|\nabla c_r\|} - k_3 \nabla c_r$$
$$\cdot\left(\nabla^T c_r(\hat{x} - x_r) + c_r - c_0\right) - k_6(x_r - \hat{x}),$$
(7.16)

where $k_6 > 0$ is a constant. As illustrated in Figure 7.3, the proposed control law consists of two parts (as shown in the dashed rectangle in the figure): the observer part and the controller part. The observer collects inputs from the robot (i.e., x_r and v_r), and inputs from the environment (i.e., c_r, ∇c_r, and $\nabla^2 c_r$)

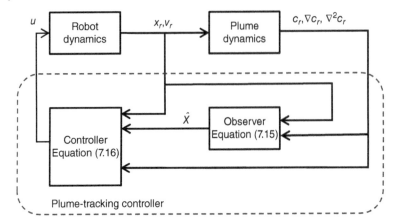

Figure 7.3 Control block diagram for the single-robot plume front tracking.

and outputs $\hat{x}$, which is the estimation on the plume front trajectory. Together with the sensory information x_r, v_r, c_r, ∇c_r, and $\nabla^2 c_r$, the observer output $\hat{x}$ is fed into the controller, which yields the control input u to the robot. The figure illustrates the interplay between the robot, the environment, and the control law.

Remark 7.5 In the control law (7.16), the term, $-k_6(x_r - \hat{x})$, forms a negative feedback on the control error. The rest of the terms are identical to $\dot{\hat{x}}$ according to the observer (7.14), and it is a velocity compensation term to regulate the tracking error.

The estimation-based control law is summarized in Algorithm 7.1.

Algorithm 7.1 Single-robot plume front tracking control

Require:
  Robot position x_r and heading θ_r, flow velocity v_r at the robot position, concentration measurement c_r, concentration gradient ∇c_r, the divergence of the gradient $\nabla^2 c_r$.
Ensure:
  Solve Problem 7.1.
1: **repeat**
2:   $x_r, \theta_r, v_r, c_r, \nabla c_r, \nabla^2 c_r \Leftarrow$ Sensor readings.
3:   update $\hat{x}$ using Equation (7.15).
4:   update u using Equation (7.16).
5:   update τ using Equation (7.3).
6: **until** time out

7.3.4 Convergence Analysis

In this part, we first present our convergence result and then prove it theoretically.

Theorem 7.1   The robot control law (7.16), under Assumptions 7.1 and 7.2, solves the single-robot plume front tracking problem defined by Problem 7.1.

For the convergence analysis, we first show the local convergence of (7.14) to systems modeled by (7.12) and (7.13), under the condition $e'(t) = \hat{x}(t) - x_r(t) = 0$. Then we show the convergence of the overall control law.

Local Convergence of Nonlinear Observer (7.14)

For the convenience of analysis, define

$$f(x) = -\frac{(v(x)^T \nabla c(x) + k \nabla^2 c(x)) \nabla c(x)}{\|\nabla c(x)\|^2} + \frac{v_d A \nabla c(x)}{\nabla c(x)}.$$

Then the system dynamics (7.12) and (7.13) can be rewritten as follows:

$$\dot{x} = f(x) \tag{7.17}$$

$$y(x) = \nabla^T c_{\hat{x}}(x - \hat{x}) + c_{\hat{x}}, \tag{7.18}$$

and the observer (7.14) can be rewritten as follows:

$$\dot{\hat{x}} = f(\hat{x}) - k_3 \nabla c_{\hat{x}}(c_{\hat{x}} - c_0). \tag{7.19}$$

The estimation error, $e(t) = x(t) - \hat{x}(t)$, has the following dynamics:

$$\dot{e} = f(x) - f(\hat{x}) + k_3 \nabla c_{\hat{x}}(c_{\hat{x}} - c_0). \tag{7.20}$$

Note that $y(x) = c_0$, and $c_{\hat{x}} - c_0$ can be approximated as

$$c_{\hat{x}} - y(x) = -\nabla^T c_{\hat{x}}(x - \hat{x}) = -\nabla^T c(x)e$$

using the first-order Taylor extension. The term, $f(x) - f(\hat{x})$, can approximated as $f(x) - f(\hat{x}) = \frac{\partial f(x)}{\partial x} e$. Note that the first-order approximation suffices local convergence analysis. Thus, the linearization of (7.20) becomes

$$\dot{e} = \frac{\partial f(x)}{\partial x} e - k_3 \nabla c(x) \nabla^T c(x) e$$

$$= \left(\frac{\partial f(x)}{\partial x} - k_3 \nabla c(x) \nabla^T c(x) \right) e. \tag{7.21}$$

Note that the x dynamic (7.17) and e dynamic (7.21) form an autonomous system. Due to the movement of the plume front and e is around 0 since we are conducting local analysis, x varies much faster than e, meaning that the averaging principle applies to (7.21) (chapter 10 in Ref. [20]). Accordingly, we get

$$\dot{e} \approx \frac{\int_{t-T}^{t} \left(\frac{\partial f(x)}{\partial x} - k_3 \nabla c(x) \nabla^T c(x) \right) dt}{T} e$$

$$= \left(\frac{\int_{t-T}^{t} \frac{\partial f(x)}{\partial x} dt}{T} - k_3 \frac{\int_{t-T}^{t} \nabla c(x) \nabla^T c(x) dt}{T} \right) e, \tag{7.22}$$

where T is the averaging length. Since $\nabla c(x)$ normally varies both in absolute value and direction with time, plus $\frac{\partial f(x)}{\partial x}$ is bounded in absolute value in a bounded area, we can expect that the symmetric matrix

$$\frac{\int_{t-T}^{t} \left(\frac{\partial f(x)}{\partial x} + (\frac{\partial f(x)}{\partial x})^T \right)/2 \, dt}{T} - k_3 \frac{\int_{t-T}^{t} \nabla c(x) \nabla^T c(x) dt}{T}$$

is negative definite by choosing a large enough k_3, which implies that $V = e^T e$ will be a proper Lyapunov function to show the convergence of e in (7.22). This completes the proof.

Convergence of the Proposed Algorithm

Substituting the control law (7.16), and noting (7.14), the robot dynamics (7.4) can be rewritten as follows,

$$\dot{x}_r = -k_6(x_r - \hat{x}) + \dot{\hat{x}} \tag{7.23}$$

Thus, the control error, $e'(t) = \hat{x}(t) - x_r(t)$, has the following dynamics:

$$\dot{e}' = -k_6 e', \tag{7.24}$$

which implies the convergence of e' to zero, that is, $\hat{x} \to x_r$ when $t \to \infty$. When time goes to infinity, the observer (7.15) reduce to Equation (7.14), which is the original extended Luenberger observer associated with the reference point movement dynamics described by (7.12) and (7.13). Such local convergence property of this nonlinear filter guarantees that $\hat{x}(t) \to x(t)$, that is, $e(t) = x(t) - \hat{x}(t) \to 0$, under the condition that $t \to \infty$ and $\hat{x}(0) \to x(0)$. Thus, $x(t) - x_r(t) = e(t) + e'(t)$ converges to zero, that is, $x_r(t)$ converges to $x(t)$. Note that $x(t)$ describes a reference point on the plume front with a patrolling speed at $v_d(t)$. Therefore, we conclude that with the control law (7.16), the robot modeled by (7.4) tracks the plume front with the concentration at c_0, and approaches a patrolling speed at $v_d(t)$ around the plume front.

7.4 Multirobot Cooperative Tracking of Plume Front

In this section, we propose a solution to the distributed multirobot plume-tracking problem (i.e., Problem 7.2) based on the results on the single-robot plume tracking. A group of robots is deployed and distributed evenly on the plume front. Like before, we still use the two-step procedure to complete the task: the estimation step and the tracking control step. In the estimation step, we define n (the number of robots in the group) moving reference points on the plume front with the desired spatial distribution, and define an observer to estimate the position of reference points recursively. In the tracking control step, we drive the robots to the estimated reference point.

For the task considered in this section, as sketched in Figure 7.1, we assign two robots as boundary robots and all the other robots as follower robots. We present the control laws for both the boundary and follower robots in Sections 7.4.1 and 7.4.2, respectively.

7.4.1 Boundary Robots

For the boundary robots, that is, robot $i = 1$ and robot $i = n$, we take the same control law as (7.16) based on a similar observer as (7.15), which writes

$$u_i = -\frac{(v_{ri}^T \nabla c_{ri} + k\nabla^2 c_{ri})\nabla c_{ri}}{\|\nabla c_{ri}\|^2} + \frac{k_{4i} A\nabla c_{ri}}{\|\nabla c_{ri}\|} - k_3\nabla c_{ri}$$

$$\cdot \left(\nabla^T c_{ri}(\hat{x}_i - x_{ri}) + c_{ri} - c_0\right) - k_6(x_{ri} - \hat{x}_i) \tag{7.25a}$$

$$\dot{\hat{x}}_i = -\frac{(v_{ri}^T \nabla c_{ri} + k\nabla^2 c_{ri})\nabla c_{ri}}{\|\nabla c_{ri}\|^2} + \frac{k_{4i} A\nabla c_{ri}}{\|\nabla c_{ri}\|} - k_3\nabla c_{ri}$$

$$\cdot \left(\nabla^T c_{ri}(\hat{x}_i - x_{ri}) + c_{ri} - c_0\right) \tag{7.25b}$$

for $i = 1, n$,

where $\hat{x}_i$ and x_{ri} are the estimated position of the ith reference point and the position of the ith robot, respectively; $v_{ri} = v(x_{ri})$, $c_{ri} = c(x_{ri})$, $\nabla c_{ri} = \nabla c(x_{ri})$, and $\nabla^2 c_{ri} = \nabla^2 c(x_{ri})$ are the flow velocity at the ith robot position x_{ri}, the concentration at x_{ri}, the gradient of the concentration at x_{ri}, and the divergence of the concentration gradient at x_{ri}, respectively; and $k_{4i} = k_{4i}(t)$ represents the desired speed along the tangent direction of the plume front.

The boundary robot control law (7.25) has an design parameter k_{4i} for $i = 1$, and $i = n$, which assigns a desired patrolling speed to the robot. By choosing $k_{41} > 0$ and $k_{4n} < 0$, the boundary robots 1 and n patrol in the counterclockwise and clockwise directions, respectively. This generates a patrolling behavior for each of the boundary robots but with an opposite direction. Combining with the follower robots whose controllers are presented in the following text, the group of robots spread out to cover the whole plume front. As the two boundary robots moves in an opposite direction along the plume front, they will meet, at which time the parameter $k_{4i}, i = 1, n$, is switched to the following one allowing them to stay together and stop rotating,

$$k_{4i} = k_7\|\hat{x}_i - \hat{x}_{n+1-i}\|\text{sign}\left((\hat{x}_i - \hat{x}_{n+1-i})^T A\nabla c(\hat{x}_i)\right), \tag{7.26}$$

where $k_7 > 0$ is a positive constant.

7.4.2 Follower Robots

We derive the follower control law based on the results for single-robot plume front tracking. The parameter $k_{4i} = k_{4i}(t)$ in (7.15), which represents the desired speed along the tangent direction, gives us an additional design freedom. This freedom is exploited in this part to design a distributed formation behavior. To this end, the gain v_d in (7.15) is replaced with an adaptive one relying on the distance difference between the two neighbors to form a cooperative observer for the reference points. Thus, we propose the following observer-based control law for the ith ($i = 2, 3, ..., n - 1$) robot:

$$u_i = -\frac{(v_{ri}^T \nabla c_{ri} + k\nabla^2 c_{ri})\nabla c_{ri}}{\|\nabla c_{ri}\|^2} + \frac{k_{4i} A\nabla c_{ri}}{\|\nabla c_{ri}\|} - k_3 \nabla c_{ri}$$
$$\cdot\left(\nabla^T c_{ri}(\hat{x}_i - x_{ri}) + c_{ri} - c_0\right) - k_6(x_{ri} - \hat{x}_i)$$

(7.27a)

$$\dot{\hat{x}}_i = -\frac{(v_{ri}^T \nabla c_{ri} + k\nabla^2 c_{ri})\nabla c_{ri}}{\|\nabla c_{ri}\|^2} + \frac{k_{4i} A\nabla c_{ri}}{\|\nabla c_{ri}\|}$$
$$-k_3\left(\nabla^T c_{ri}(\hat{x}_i - x_{ri}) + c_{ri} - c_0\right)\nabla c_{ri}$$

(7.27b)

with

$$k_{4i} = \frac{k_5(\hat{x}_{i-1}^T A\nabla c_{ri-1} + \hat{x}_{i+1}^T A\nabla c_{ri+1} - 2\hat{x}_i^T A\nabla c_{ri})}{\|A\nabla c_{ri}\|}$$

for $i = 2, 3, ..., n-1,$

(7.27c)

where $k_5 > 0$ is a constant.

The block diagram of the multirobot cooperative plume front tracking for a follower robot is illustrated in Figure 7.4. In addition to the information from the environment and the ith robot itself, this control scheme also requires information of $\hat{x}_{i-1}$ and $\hat{x}_{i+1}$, which are the observer outputs from the $i - 1$th and the ith robot, respectively.

The procedure of the proposed algorithm to solve Problem 7.2 is stated in Algorithm 7.2.

7.4.3 Convergence Analysis

In this section, we first present our convergence result and then prove it theoretically.

Theorem 7.2 The boundary robot control law (7.25) together with the follower robot control law (7.27), under Assumptions 7.1 and 7.2,

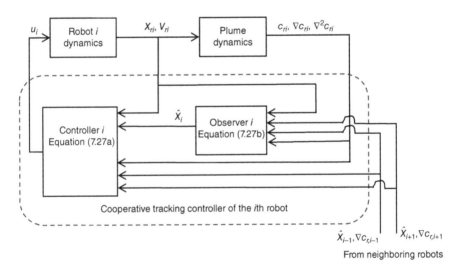

Figure 7.4 Control block diagram of a follower robot in the multirobot plume front tracking.

Algorithm 7.2 Multirobot cooperative plume front tracking for robot i (boundary robots: $i = 1, n$; follower robots: $i = 2, 3, ..., n - 1$)

Require:
For boundary robots $i = 1, n$: robot position x_{ri} and heading θ_{ri}, flow velocity v_i at the robot position, concentration measurement c_{ri}, concentration gradient ∇c_{ri}, the divergence of the gradient $\nabla^2 c_{ri}$, the desired patrolling speed k_{4i}.
For follower robots $i = 2, 3, ..., n - 1$: robot position $x_{ri-1}, x_{ri}, x_{ri+1}$, robot heading θ_{ri}, flow velocity $v_{ri-1}, v_{ri}, v_{ri+1}$ concentration measurement c_{ri-1}, c_{ri}, c_{ri+1} concentration gradient $\nabla c_{ri-1}, \nabla c_{ri}, \nabla c_{ri+1}$, the divergence of the gradient $\nabla^2 c_{ri-1}, \nabla^2 c_{ri}, \nabla^2 c_{ri+1}$.

Ensure:
Solve Problem 7.2.
1: **repeat**
2:   **if** $i = 1, n$ **then**
3:     $c_{ri}, \nabla c_{ri}, \nabla^2 c_{ri} \Leftarrow$ Sensor readings.
4:     update $\hat{x}_i$ using Equation (7.25b).
5:     update u_i using Equation (7.25a).
6:     update τ_i using Equation (7.3).
7:   **else**
8:     $v_r, c_r, \nabla c_r, \nabla^2 c_r \Leftarrow$ Sensor readings.
9:     $c_{r-1}, \nabla c_{r-1} \Leftarrow$ Communication with Robot $i - 1$.
10:     $c_{r+1}, \nabla c_{r+1} \Leftarrow$ Communication with Robot $i + 1$.
11:     update k_{4i} using Equation (7.27c).
12:     update $\hat{x}_i$ using Equation (7.27b).
13:     update u_i using Equation (7.27a).
14:     update τ_i using Equation (7.3).
15:   **end if**
16: **until** time out

solves the distributed multirobot plume front tracking problem defined by Problem 7.2.

The convergence of the boundary robots to the plume front can be similarly analyzed as the convergence analysis in the single-robot case. We now turn to the convergence analysis of the follower robots toward even distribution on the plume front between boundary robots. For analysis convenience of the follower robots, we first construct the following cooperative reference points on the plume front for the follower robots labeled as $i = 2, 3, ..., n - 1$:

$$\dot{x}_i = -\frac{(v_{ri}^T \nabla c_{ri} + k\nabla^2 c_{ri}) \nabla c_{ri}}{\|\nabla c_{ri}\|^2} + \frac{k_{4i} A \nabla c_{ri}}{\|\nabla c_{ri}\|} \tag{7.28}$$

$$y(x_i) = \nabla^T c_{ri}(x_i - x_{ri}) + c_{ri} \tag{7.29}$$

with k_{4i} defined in (7.27c). By comparing the single-robot control law (7.16) and the follower control law (7.27a) in the multirobot case, we find the follower control law falls in a special case of the single-robot control law by assigning $v_d(t)$ in (7.16) with a specific value defined in (7.27c). Therefore, the results drawn based on the more general case for the control law (7.16) apply to the special case with the control law (7.27a), that is, the control law (7.27a) drives the follower robots to the reference points on plume front modeled by (7.28), that is, $\hat{x}_i \to x_i$ when $t \to \infty$ for $i = 2, 3, ..., n - 1$.

Now, we turn to analyze the impact of (7.27c) for the ultimate formation of the robot group. When $t \to \infty$, we have $\hat{x}_i = x_{ri} = x_i$ and further have the following by left multiplying $(A\nabla c_i)^T = (A\nabla c(x_i))^T$ on both sides of (7.28) with noting the $\hat{x}_i \to x_i$,

$$(A\nabla c_i)^T \dot{x}_i = k_{4i}\|A\nabla c_i\| = k_5\big((A\nabla c_{i-1})^T$$
$$\cdot x_{i-1} + (A\nabla c_{i+1})^T x_{i+1} - 2(A\nabla c_i)^T x_i\big). \tag{7.30}$$

As $(A\nabla c(x_i))^T \dot{x}_i \approx \frac{\partial(A\nabla c_i)^T x_i)}{\partial t}$ provided that the changing rate of $\nabla c(x_i)$ is small, we thus result in the following by defining a new metric $z_i = (A\nabla c_i)^T x_i$:

$$\dot{z}_i = k_5(z_{i-1} - z_i + z_{i+1} - z_i). \tag{7.31}$$

Clearly, this is a consensus protocol [17] with two boundary robots $z_1 = (A\nabla c_1)^T x_1$ and $z_n = (A\nabla c_n)^T x_n$. It can be drawn that z_i converges to $z_1 + \frac{(i-1)(z_n-z_1)}{n-1}$ for $i = 2, 3, ..., n - 1$. This yields that $z_{i+1} - z_i = z_i - z_{i-1}$ for all $i = 2, 3, ..., n - 1$ and implies that the neighboring robots has the same distance between each other along the tangent direction of the plume front (note that $A\nabla c_i$ represents the tangent direction at the ith robot along the plume front), which achieves the even distribution objective along the plume front between the two boundary robots.

7.5 Simulations

In this section, we present numerical simulation results using MATLAB to validate the effectiveness of the proposed algorithms.

7.5.1 Simulation Environment

The simulation of the environment includes two parts: one is the simulator of a flow field; the other is the simulator of the concentration field. The flow field is generated numerically by solving incompressible Navier–Stokes equations in a rectangular domain with prescribed velocities along the boundary using the program presented in Ref. [21]. The concentration field is generated usingfinite difference method to numerically solve the partial differential equation (7.1) with a fixed value of the concentration at sources as boundary conditions. The flow field is visualized in Figure 7.5 (a), where the strength and direction of the flow at various positions are shown. Both of the two chemical sources locating at $(13, 3)$ and $(7, 7)$ have a fixed concentration 3. The propagation of the chemical in the flow field results in a time-varying concentration field.

7.5.2 Single-Robot Plume Front Tracking

Here, we validate Algorithm 7.1 that uses a single robot to tack and patrol the plume front with concentration value $c_0 = 0.3$. The diffusion coefficient k is set as $k = 0.5$. Due to the advection effect, the chemicals are blown along the flow direction and form a nonsymmetric contour in the plume front. In the simulation, the patrolling speed is set as $k_4 = 6$, the gradient gain is set as $k_3 = 5$, and the estimated reference tracking gain is set as $k_6 = 5$. The robot is deployed to the field at time $t = 2$ s, after the sources starts propagation at time $t = 0$ s. As shown in Figure 7.5 (b), due to the chemical propagation, the plume front contour expands with time. With the proposed method, the robot adaptively adjusts its orbits to track the plume front. We denote the difference between the plume front concentration c_0 and the measured concentration at the robot position c_r as the plume front tracking error. As shown in Figure 7.5 (c), the error attenuates to zero along time. Note that the lug in the error curve in Figure 7.5(c) around time $t = 8.5$ s happens at the time when the robot travels across the lower right tip of the plume front curve, where the concentration changes sharply since the area is mostly close to the source and locates in a place with the flow direction adverse to the diffusion direction. Any small variation in the robot trajectory in this area results in a large error in the concentration values.

7.5.3 Multirobot Cooperative Plume Front Tracking

We validate Algorithm 7.2 with 30 robots to cover the dynamic plume front with concentration $c_0 = 0.1$. The diffusion coefficient k in (7.1) is set as $k = 0.5$. For the boundary robots, the patrolling speed is set as $k_{4i} = 3$ for $i = 1$ and $k_{4i} = -3$ for $i = 30$. For follower robots, the gain k_5 is set to be $k_5 = 5$. For all robots, the gradient gain is set as $k_3 = 5$ and the estimated reference tracking gain is set as $k_6 = 5$. After the sources start propagation for 2 s, 30 robots are deployed as shown in Figure 7.6, where the follower robots are marked by hollow squares and the boundary robots are marked by hollow circles. We want the robots to distribute on the plume front evenly, move with the plume propagation and expand the formation when the plume front expands. For this purpose, starting from time $t = 6s$, when the condition $\|\hat{x}_1 - \hat{x}_{30}\| \leq l_{dis}$ with $l_{dis} = 2$ is satisfied, k_{4i} switches to the control law (7.25) with the parameter k_{4i} set as (7.26) for the boundary robots $i = 1$ and $i = 30$, where the gain k_7 is set to be 20 in the simulation.

Figure 7.6 shows the snapshots of the simulation results along time. After the robots are deployed in the field, it rapidly forms the desired even distribution

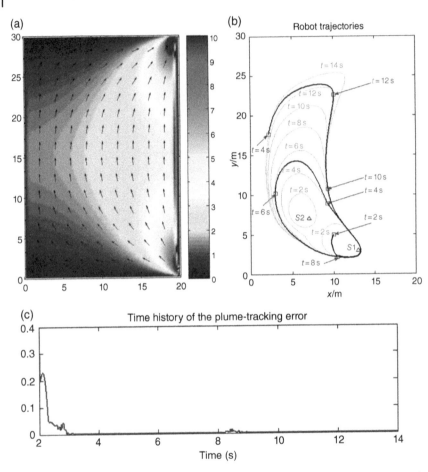

Figure 7.5 (a) The velocity field considered in the simulation, where the pseudo-color and the yellow arrow indicate the strength and the direction of the flow, respectively. (b) The robot trajectory for single-robot plume front tracking and patrolling, where S_1 and S_2 are the two sources; the green curves and the red squares represent the plume front contour and the robot positions, respectively; and black curve represents the robot trajectory. (c) The plume front tracking error. (*See insert for color representation of the figure.*)

and reach the plume front as shown at time $t = 4$ s. As time elapses, the robots move around the plume front shown as the black contour, and simultaneously follow the expansion movement of the plume front as shown in the snapshot $t = 6$ s. After this time, the boundary robots keep moving in an opposite direction to cover the plume front until the distance between them is within the value l_{dis}, which is satisfied at time $t = 10$ s. Finally, as shown at $t = 12$ s, the two boundary robots reach consensus in their positions and the follower robots are successfully deployed evenly covering the plume front.

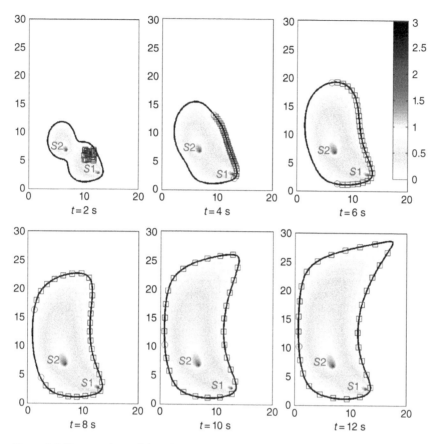

Figure 7.6 The snapshots of the robot movements for a typical simulation run with 30 robots for cooperative plume front tracking, where the pseudo-color indicates the concentration distribution at each time step with the scale shown in the colorbar aside; the black curve is the contour of the plume front with a concentration value $c_0 = 0.01$; the hollow square in blue and the hollow circle in red represent the position of the follower robots and the true position of the boundary robots, respectively; and the positions marked with S1 and S2 are the two sources starting propagating chemicals in the flow field shown in Figure 7.5 at time $t = 0$. (*See insert for color representation of the figure.*)

To evaluate the performance, we define two indices: one is the formation error and the other is the plume-tracking error. The formation error measures whether the formation is achieved for the follower robots, $i = 2, 3, ..., n - 1$. As the distance between any two neighboring robots is $\left| \|x_{ri} - x_{r,i+1}\| - \|x_{ri} - x_{r,i-1}\| \right|$, we define the average distance of the group of $(n - 2)$ robots as the group formation error: $\sqrt{\frac{\sum_{i=2}^{n-1} (\|x_{ri} - x_{r,i+1}\| - \|x_{ri} - x_{r,i-1}\|)^2}{n-2}}$. We define the plume-tracking error of the group to be $\sqrt{\frac{\sum_{i=1}^{n} (c(x_{ri}) - c_0)^2}{n}}$, which is

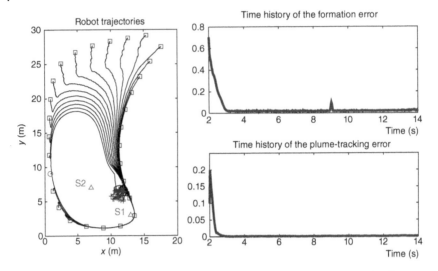

Figure 7.7 The robot trajectories, the time profile of the formation error, and the plume front tracking error, where S_1 and S_2 represent the sources, "+" represents the initial position of robots, and the hollow squares and the hollow circles represent the final positions of the follower robots and the boundary robots, respectively.

the average tracking error of each robot in the group. The time histories of the formation error and the plume-tracking error are plotted in Figure 7.7, where the trajectories of the robots are also plotted. It can be seen that both the formation error and the tracking error reduce rapidly to a small value. It also shows that a small peak appears at around $t = 10$ s in the formation error subfigure, which is caused by the switching of the boundary robots' controllers from spread-out to stay-close and stop-rotating. It can be observed that the performance is overall satisfactory.

7.6 Conclusion

In this chapter, the dynamic plume-tracking problem was solved using a group of cooperating robots. The advection–diffusion equation was used to model the pollution plume propagation, and a model-based observer was built to estimate the plume front trajectory. Solutions were then provided for single-robot tracking and multirobot tracking with simultaneous formation based on an estimation and control framework. The multirobot plume tracking is distributed and scalable with a nearest-neighboring communication topology. Simulation results showed effectiveness of the proposed strategy.

Notes

The results in this chapter were first published in Ref. [22].

Acknowledgment is given to ©2014 IEEE. Reprinted, with permission, from Li, S., Guo, Y., and Bingham, B. (2014) Multi-robot cooperative control for monitoring and tracking dynamic plumes, in IEEE International Conference on Robotics and Automation, pp. 67–73.

References

1 Lilienthal, A. and Duckett, T. (2004) Building gas concentration gridmaps with a mobile robot. *Robotics and Autonomous Systems*, **48**, 3–16.

2 Farrell, J., Pang, S., and Li, W. (2003) Plume mapping via hidden markov methods. *IEEE Transactions on Systems, Man, and Cybernetics, Part B: Cybernetics*, **33** (6), 850–863.

3 Pang, S. and Farrell, J. (2006) Chemical plume source localization. *IEEE Transactions on Systems, Man, and Cybernetics, Part B: Cybernetics*, **36** (5), 1068–1080.

4 Ferri, G., Jakuba, M., Mondini, A., Mattoli, V., Mazzolai, B., Yoerger, D., and Dario, P. (2011) Mapping multiple gas/odor sources in an uncontrolled indoor environment using a bayesian occupancy grid mapping based method. *Robotics and Autonomous Systems*, **59** (11), 988–1000.

5 Li, W., Farrell, J., Pang, S., and Arrieta, R. (2006) Moth-inspired chemical plume tracing on an autonomous underwater vehicle. *IEEE Transactions on Robotics*, **22** (2), 292–307.

6 Bruemmer, D.J., Dudenhoeffer, D.D., McKay, M.D., and Anderson, M.O. (2002) A robotic swarm for spill finding and perimeter formation, in *Spectrum*, Reno, NV, pp. 28–35.

7 Lilienthal, A. and Duckett, T. (2003) Experimental analysis of smelling braitenberg vehicles. *Environment*, **5**, 1–10.

8 Clark, J. and Fierro, R. (2005) Cooperative hybrid control of robotic sensors for perimeter detection and tracking, in *American Control Conference*, Portland, OR, vol. 5, pp. 3500–3505.

9 Hsieh, M., Forgoston, E., Mather, T., and Schwartz, I. (2012) Robotic manifold tracking of coherent structures in flows, in *IEEE International Conference on Robotics and Automation*, pp. 4242–4247, doi:10.1109/ICRA.2012.6224769.

10 Zhang, F., Fiorelli, E., and Leonard, N. (2007) Exploring scalar fields using multiple sensor platforms: Tracking level curves, in *IEEE Conference on Decision and Control*, pp. 3579–3584, doi:10.1109/CDC.2007.4434245.

11 Malisoff, M. and Zhang, F. (2013) Adaptive control for planar curve tracking under controller uncertainty. *Automatica*, **49** (5), 1411–1418.

12 Wu, W. and Zhang, F. (2011) Cooperative exploration of level surfaces of three dimensional scalar fields. *Automatica*, **47** (9), 2044–2051.

13 Smith, R., Yi, C., Li, P., Caron, D., Jones, B., and Sukhatme, G. (2010) Planning and implementing trajectories for autonomous underwater vehicles to track evolving ocean processes based on predictions from a regional ocean model. *International Journal of Robotic Research*, **29** (12), 1475–1497.

14 Sonnenburg, C. and Woolsey, C. (2013) Modeling, identification, and control of an unmanned surface vehicle. *Journal of Field Robotics*, **30** (3), 371–398.

15 Biyik, E. and Arcak, M. (2007) Gradient climbing in formation via extremum seeking and passivity-based coordination rules, in *46th IEEE Conference on Decision and Control*, New Orleans, LA, vol. 3, pp. 3133–3138.

16 Henderson, T. and Grant, E. (2004) Gradient calculation in sensor networks, in *IEEE/RSJ International Conference on Intelligent Robots and Systems*, Sendai, Japan, vol. 2, pp. 1792–1795.

17 Saber, R., Fax, J., and Murray, R. (2007) Consensus and cooperation in networked multi-agent systems. *Proceedings of the IEEE*, **95**, 215–233.

18 Klavins, E. (2003) Communication complexity of multi-robot systems, in *Algorithmic Foundations of Robotics V*, vol. 7 (eds J. Boissonnat, J. Burdick, K. Goldberg, and S. Hutchinson), Springer, Berlin/Heidelberg.

19 Zeitz, M. (1987) The extended luenberger observer for nonlinear systems. *Systems and Control Letters*, **9** (2), 149–156.

20 Khalil, H. (2002) *Nonlinear Systems*, Prentice Hall, Upper Saddle River, NJ.

21 Seibold, B. (2008) A compact and fast matlab code solving the incompressible navier-stokes equations on rectangular domains. Online at http://math.mit.edu/~gs/cse/codes/mit18086_navierstokes.pdf, Accessed 6 October 2016.

22 Li, S., Guo, Y., and Bingham, B. (2014) Multi-robot cooperative control for monitoring and tracking dynamic plumes, in *IEEE International Conference on Robotics and Automation*, pp. 67–73, doi:10.1109/ICRA.2014.6906591.

Part III

Distributed Cooperative Control for Multiagent Physics Systems

Part III considers applications of cooperative control to multiagent physics systems. Such class of systems are different from the networked communication systems and the robotic systems that are discussed in Parts I and II. Due to intrinsic physics principles, multiagent physics systems exhibit complex behaviors that are highly nonlinear. There may be limited accessibility of system elements, and conventional control tools may not be feasible. This part of the book addresses challenging cooperative control problems in two representative physics systems, both of which share common features including (i) the coupling structure of system elements can be described by graph Laplacians reviewed earlier in the book, and (ii) the nonlinear coupling of system elements plays an important role in complex system behaviors.

Chapter 8 presents friction control of coupled nanoparticle arrays. Friction and wear is estimated to cost the US economy 6% of the gross national product. The study of nanoscale friction has technological impacts in reducing energy loss in machines, in microelectromechanical systems (MEMS), and in the development of durable and low-friction surfaces and ultrathin lubrication films. A majority of the work on friction control at the nanoscale from the physics community study physics principles of frictional phenomena with experimental and/or numerical simulation support. Analytic studies of the dynamic model, rigorous validations of control techniques, and theoretical guidelines on how to quantitatively adjust tunable control variables or parameters are missing. This chapter presents analytic control methods for a one-dimensional (1D) coupled particle array that are modeled by the nonlinear Frenkel–Kontorova (FK) model, and the control objective is to achieve smooth sliding of the particle array on a substrate thus to reduce friction. Nonlinear particle interactions are considered, matrix theory on graph Laplacians is used in system stability analysis, and control limitations on nanoscale systems are also taken into consideration in the control design.

Chapter 9 presents coupled semiconductor lasers and its synchronization control. From all the lasers, semiconductor lasers are considered to be mostly

Distributed Cooperative Control: Emerging Applications, First Edition. Yi Guo.
© 2017 John Wiley & Sons, Inc. Published 2017 by John Wiley & Sons, Inc.
Companion website: www.wiley.com/go/Guo/DistributedCooperativeControl

energy efficient with electro-optical efficiency nearing to 80% (e.g., solid-state lasers are 15% electro-optically efficient and fiber lasers are 40%). Semiconductor lasers are also mostly compact and less expensive. All these qualities make semiconductor lasers very attractive candidates for numerous energy-related applications. The major challenge though is achieving excellent beam quality and coherence from large arrays of semiconductor lasers. A single high-power semiconductor laser can provide approximately 1–2 W of power. Combining such lasers into arrays may result in much higher powers (at the level of KWs) and enable wide variety of applications in the energy efficient and economical way. This chapter considers the coupled Lang and Kobayashi model of semiconductor laser arrays, analyzes synchronizing behaviors, and characterizes coupling conditions for synchronization using graph Laplacian matrix tools.

8

Friction Control of Nano-particle Array

8.1 Introduction

Tribology has been an active research area due to its broad applications in the fields of physics, chemistry, geology, biology, and engineering [1]. Rapidly growing areas of tribology are in microelectromechanical systems (MEMS) and biological systems, particularly the lubrication mechanisms in joints. Recent advances have substantially improved the understanding of frictional phenomena, particularly on the inherently nonlinear nature of friction [2]. Traditionally, the control of frictional forces has been approached by chemical means, such as supplementing base lubricants with friction modifier additives. A recent different approach, which tunes frictional responses by controlling the system mechanically via normal vibrations of small amplitude and energy, has attracted considerable interest, see Refs. [3–8]. The idea is to reduce the friction force or to eliminate stick-slip motion through a stabilization of desirable modes of motion. We follow this line of research and design feedback control laws to control frictional dynamics toward a desirable mode of motion.

Friction can be manipulated by applying perturbations to accessible elements and parameters of a sliding system [1]. The authors [3] proposed an intriguing idea to control the overall motion of an array of mechanically coupled objects sliding on a dissipative substrate via feedback control and tracking, and applied the idea to a particle array, with the frictional dynamics described by the Frenkel–Kontorova (FK) model. A control problem was formulated therein, and a global feedback control scheme was presented to render the system's output, the velocity of the center of mass of the nanoarray, to approach a given targeted value, subject to some fluctuations. Results were supported by simulations only. Theoretical justification on the non-Lipschitzian control was later given in Ref. [9]. However, we showed in Ref. [10] that the control law in Ref. [3] does not eliminate the persistent oscillations of the controlled variables around their equilibrium points.

We study in this chapter the problem of controlling frictional dynamics of a one-dimensional (1D) particle array using control theoretical methods. We describe the FK model to characterize the dynamics of the interconnected 1D

Distributed Cooperative Control: Emerging Applications, First Edition. Yi Guo.
© 2017 John Wiley & Sons, Inc. Published 2017 by John Wiley & Sons, Inc.
Companion website: www.wiley.com/go/Guo/DistributedCooperativeControl

particle system. It is a nonlinear system since both the coupling of the particles with the substrate and the particle interactions are nonlinear. A control problem is then formulated based on the FK model, which is a constrained nonlinear control problem. The constraint is caused by the inaccessibility of individual particles. The control objective is to achieve tracking of the targeted velocity using physically accessible variables, that is, the average quantity of the interconnected system. We present two main results in the chapter. First, we study stability of equilibrium points of the particle array in the presence of linear and nonlinear interparticle coupling, respectively. Second, we design global tracking control laws to achieve that the average velocity of the array, that is, the velocity of the center mass, tracks any given constant targeted velocity. Global feedback control laws are explicitly constructed using the Lyapunov theory–based method. We further analyze local stability of individual particles in the closed-loop system under the average control law. Finally, we illustrate the control performances using MATLAB simulations of different sizes of a particle array.

Organization: The rest of the chapter is organized as follows: Section 8.2 presents the FK model used to describe the frictional dynamics. In Section 8.3, the local stability of the open-loop interconnected particle system is analyzed in two subsections with linear and nonlinear particle interactions, respectively. Then, a tracking control problem is defined in Section 8.4. Section 8.5 presents tracking control design to solve the control problem formulated in Section 8.4. Simulations results are given in Section 8.6. The chapter is finally concluded with brief remarks in Section 8.7.

8.2   The Frenkel–Kontorova Model

The basic equations for the driven dynamics of a 1D particle array of N identical particles moving on a surface are given by a set of coupled nonlinear equations [3, 11] as follows:

$$m' \ddot{z}_i + \gamma' \dot{z}_i = -\frac{\partial U(z_i)}{\partial z_i} - \sum_{\substack{j=1 \\ j \neq i}}^{N} \frac{\partial W(z_i - z_j)}{\partial z_i} + f_i' + \eta(t). \qquad (8.1)$$

Here, $i = 1, \dots, N$, z_i is the coordinate of the ith particle, m' is its mass, γ' is the positive friction coefficient representing the single particle energy exchange with the substrate, f_i' is the applied external force, $\eta(t)$ denotes additive the Gaussian noise, $U(z_i)$ is the periodic potential applied by the substrate, and $W(z_i - z_j)$ is the interparticle interaction potential.

Under the simplifications that the substrate potential is in the form of $(m'/4\pi^2)(1 - \cos 2\pi z_i)$, the same force is applied to each particle, and there is zero noise (i.e., $\eta(t) = 0$), the equation of motion reduces to the following FK model:

$$\ddot{\phi}_i + \gamma \dot{\phi}_i + \sin(\phi_i) = f + F_i, \qquad (8.2)$$

where ϕ_i is the dimensionless phase variable, $\phi_i = 2\pi z_i$,

$$\gamma = \gamma'/m', f = 2\pi f_i'/m',$$

$$F_i = \frac{2\pi}{m'} \sum_{j=1, j\neq i}^{N} \frac{\partial W(\phi_i/2\pi - \phi_j/2\pi)}{\partial(\phi_i/2\pi)}.$$

A specific example often considered for the particle interaction force, F_i, is the nearest-neighbor interaction in the form of Morse-type interaction [3, 11]:

$$F_i = \frac{\kappa}{\beta} \left\{ e^{-\beta(\phi_{i+1}-\phi_i)} - e^{-2\beta(\phi_{i+1}-\phi_i)} \right\}$$

$$- \frac{\kappa}{\beta} \left\{ e^{-\beta(\phi_i-\phi_{i-1})} - e^{-2\beta(\phi_i-\phi_{i-1})} \right\},$$

$$i = 2, \dots, N-1, \tag{8.3}$$

where κ and β are positive constants. The free-end boundary conditions are represented as follows:

$$F_1 = \frac{\kappa}{\beta} \left\{ e^{-\beta(\phi_2-\phi_1)} - e^{-2\beta(\phi_2-\phi_1)} \right\},$$

$$F_N = -\frac{\kappa}{\beta} \left\{ e^{-\beta(\phi_N-\phi_{N-1})} - e^{-2\beta(\phi_N-\phi_{N-1})} \right\}. \tag{8.4}$$

As $\beta \to 0$, (8.3) turns to

$$F_i = \kappa \left(\phi_{i+1} - 2\phi_i + \phi_{i-1} \right), \quad i = 2, \dots, N-1, \tag{8.5}$$

which represents a linear approximation of particle interaction for small β with the following free-end boundary conditions:

$$F_1 = \kappa(\phi_2 - \phi_1), \quad F_N = \kappa(\phi_{N-1} - \phi_N). \tag{8.6}$$

An illustration of the FK model is shown in Figure 8.1.

The FK model (8.2) describes a chain of particles interacting with the nearest neighbors in the presence of an external periodic potential. It is one of the best known simple models for frictional dynamics, and can be extended to 2D and 3D models and to a full set of molecular dynamics. Besides describing the friction dynamics [1], the FK model has been widely invoked in descriptions of

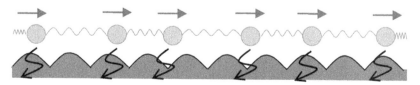

Figure 8.1 The Frenkel–Kontorova model represents a harmonic chain (which mimics a layer of nanoparticles) in a spatially periodic potential (which mimics the substrate). The chain is driven by a constant force which is damped by a velocity-proportional damping.

many other physical problems, such as charge-density waves, magnetic spirals, and absorbed monolayers [12].

The FK model presents a nonlinear interconnected systems. Nonlinearity appears since (i) the coupling of the particles with the substrate is nonlinear and (ii) the particle interaction is nonlinear. Particularly, the nonlinear Morse-type interaction represents an attraction force between two nearest particles when their distance is longer than the natural length of the spring, and a restoring force (increasing unlimited) between them when the distance is shorter than the natural length of the spring [13]. The Morse-type particle interaction presents a class of attraction/repulsive functions for a 1D swarm aggregation, which may be of interest to the research in swarm dynamics, see Refs. [14] and [15].

8.3 Open-Loop Stability Analysis

Before we define our control problem, we study the stability of the open-loop system of the FK model (8.2). The dynamics in (8.2) expressed without external forces can be equivalently written as

$$
\dot{x}_{i1} = x_{i2}
$$
$$
\dot{x}_{i2} = -\sin x_{i1} - \gamma x_{i2} + F_i, \tag{8.7}
$$

where $i = 1, 2 \ldots, N, x_{i1} = \phi_i, x_{i2} = \dot{\phi}_i$, and F_i is the Morse-type particle interaction. Let's look at the local stability of the equilibrium points in the presence of (i) linear particle interaction and (ii) nonlinear particle interaction, respectively.

8.3.1 Linear Particle Interactions

We consider the local stability of (8.7) when F_i takes the form of the linear interaction given in (8.5). From (8.7), the equilibrium points are at $(x_{i1}, x_{i2}) = (x_{i1}^*, 0)$ where x_{i1}^* are solutions to

$$
-\sin x_{11}^* + \kappa(x_{21}^* - x_{11}^*) = 0,
$$
$$
-\sin x_{i1}^* + \kappa(x_{i+1,1}^* - 2x_{i1}^* + x_{i-1,1}^*) = 0, i = 2, \ldots, N-1,
$$
$$
-\sin x_{N1}^* + \kappa(x_{N-1,1}^* - x_{N1}^*) = 0. \tag{8.8}
$$

Define new states as $z_{i1} = x_{i1} - x_{i1}^*, z_{i2} = x_{i2}$, and linearize it around its equilibrium. We obtain

$$
\dot{z}_{i1} = z_{i2}
$$
$$
\dot{z}_{i2} = -\cos x_{i1}^* z_{i1} - \gamma z_{i2} + \kappa(z_{i+1,1} - 2z_{i1} + z_{i-1,1})
$$
$$
\quad - \sin x_{i1}^* + \kappa(x_{i+1,1}^* - 2x_{i1}^* + x_{i-1,1}^*)
$$
$$
= -\cos x_{i1}^* z_{i1} - \gamma z_{i2} + \kappa(z_{i+1,1} - 2z_{i1} + z_{i-1,1}). \tag{8.9}
$$

Note that the last Equal sign holds due to the equilibrium Equation (8.8).

Stacking the state space equations for $i = 1, 2, \ldots, N$, we obtain

$$\dot{z} = \overline{A}z + \overline{B}\,\overline{Q}z \qquad (8.10)$$

where $z = [z_{11}, z_{12}, z_{21}\ z_{22}, \ \ldots \ z_{N1}, z_{N2}]^T$,

$$\overline{A} = I_N \otimes A, \overline{B} = I_N \otimes B, \overline{Q} = Q \otimes \begin{bmatrix} 1 & 0 \end{bmatrix}, \qquad (8.11)$$

and

$$A = \begin{bmatrix} 0 & 1 \\ 0 & -\gamma \end{bmatrix}, \ B = \begin{bmatrix} 0 \\ 1 \end{bmatrix}, \qquad (8.12)$$

$$Q =
\begin{bmatrix}
-\kappa - \cos x_{11}^* & \kappa & 0 & \cdots & & 0 \\
\kappa & -2\kappa - \cos x_{21}^* & \kappa & 0 & & \cdots \\
& & \vdots & & & \\
0 & \cdots & \kappa & -2\kappa - \cos x_{N-1,1}^* & \kappa \\
0 & \cdots & 0 & \kappa & -\kappa - \cos x_{N1}^*
\end{bmatrix}$$

$$\in \mathfrak{R}^{N \times N}. \qquad (8.13)$$

Note that $\overline{A}$ and $\overline{Q}$ can be represented by other matrices. We represented them as the current form for the convenience of the proof of Theorem 8.1 presented in the following.

Theorem 8.1 The system (8.7) with linear particle interaction (8.5) is locally asymptotically stable at the equilibrium points $(x_{i1}^*, 0)$ if all of the eigenvalues of the matrix Q defined in (8.13) have negative real parts; it is unstable if any of the eigenvalues of the matrix Q has a positive real part. Particularly, it is locally asymptotically stable if $\cos x_{i1}^* \geq 0$ for all i with strict inequality for atleast one i, and it is unstable if $\cos x_{i1}^* \leq 0$ for all i with strict inequality for at least one i.

Next, we present the proof of Theorem 8.1, which needs Lemmas 8.1–8.3.

Lemma 8.1 ([16], page 171) **Spectral Theorem for Symmetric Matrices:** If A is an $n \times n$ real symmetric matrix, then there always exist matrices L and D such that $L^T L = LL^T = I$ and $LAL^T = D$, where D is the diagonal matrix of eigenvalues of A ([16], page 171).

Lemma 8.2 ([17]) Let $A = [a_{ij}]_{i,j=1}^n \in \mathfrak{R}^{n \times n}$ and assume that $a_{ii} > 0$ for each i and $a_{ij} \leq 0$ whenever $i \neq j$. If A is diagonally dominant, that is,

$$a_{ii} > \sum_{j=1, j \neq i}^{n} |a_{ij}|, \quad i = 1, 2, \ldots, n,$$

or, if A is irreducible and

$$a_{ii} \geq \sum_{j=1, j\neq i}^{n} |a_{ij}|, \quad i = 1, 2, \ldots, n,$$

with strict inequality for at least one i, then A is an M-matrix. A symmetric M-matrix is positive definite.

Lemma 8.3 *([18], Appendix A)*   Define the set W consisting of all zero row sum matrices which have only nonpositive off-diagonal elements. A matrix $A \in W$ satisfies the following:

1. All eigenvalues of A are nonnegative.
2. 0 is an eigenvalue of A.
3. 0 is an eigenvalue of multiplicity 1 if A is irreducible.

Proof of Theorem 8.1: First, we study stability of the linearized system (8.10) for any positive constants γ, κ and for any $N \geq 2$. To perform this stability analysis, we find a transformation matrix to transform the system matrix into a block diagonal one.

Define a similarity transformation $z = \overline{T}\zeta$. In the new coordinate, the system dynamics is

$$\dot{\zeta} = H\zeta. \tag{8.14}$$

We show how to choose $\overline{T}$ and present H accordingly.

Since Q is a real symmetric matrix, according to Lemma 8.1, there exists a unitary matrix T such that $T^{-1}QT = D$ where D is a diagonal matrix of eigenvalues of Q. Let

$$\overline{T} = T \otimes I_2 \tag{8.15}$$

where I_2 is the 2×2 identity matrix. Then

$$
\begin{aligned}
H &= \overline{T}^{-1}(\overline{A} + \overline{B}\,\overline{Q})\overline{T} \\
&= \overline{T}^{-1}\left[I_N \otimes A + (I_N \otimes B)(Q \otimes [\,1\ 0\,])\right]\overline{T} \\
&= \overline{T}^{-1}\left(I_N \otimes A + Q \otimes \begin{bmatrix} 0 & 0 \\ 1 & 0 \end{bmatrix}\right)\overline{T} \\
&= \left(T^{-1}I_N T\right) \otimes A + \left(T^{-1}QT\right) \otimes \begin{bmatrix} 0 & 0 \\ 1 & 0 \end{bmatrix} \\
&= I_N \otimes A + D \otimes \begin{bmatrix} 0 & 0 \\ 1 & 0 \end{bmatrix}. \tag{8.16}
\end{aligned}
$$

We can see that H is block diagonal, and the block diagonal element of H writes

$$H_{ii} = \begin{bmatrix} 0 & 1 \\ \alpha_i & -\gamma \end{bmatrix}, \tag{8.17}$$

where $\alpha_i, i = 1, 2, \ldots, N$ are eigenvalues of Q. The stability of the system depends on the sign of the real parts of $\alpha_i, i = 1, \ldots, N$.

1. If $\alpha_i, i = 1, 2, \ldots, N$ have negative real parts, the eigenvalues of $H_{ii}, i = 1, 2, \ldots, N$ have also negative real parts, and so does the matrix H. This indicates that the system is asymptotically stable at these points. Due to the similarity transformation, the same stability result holds for the original system $\dot{z} = (\overline{A} + \overline{B}\,\overline{Q})z$. Further, local stability of the original nonlinear system (8.7) can be deduced from the stability analysis of its linearized system (8.10) ([19], Theorem 3.1).

2. If α_i has a positive real part for any $i \in [1, N]$, eigenvalues of $H_{ii}, i = 1, 2, \ldots, N$, also have positive real parts. With the same arguments as before, the system (8.7) is unstable at these points.

Checking the structure of matrix Q in (8.13), we have the following cases:

- If $\cos x_{i1}^* \geq 0$ for all i with strict inequality for at least one i, the matrix $-Q$ is an M-matrix and $\alpha_i < 0$ for all i according to Lemma 8.2. Therefore, Q is Hurwitz and the system is asymptotically stable.
- If $\cos x_{i1}^* = 0$ for all i, Q has one (and only one) eigenvalue 0 according to Lemma 8.3. The linear system (8.10) is marginally stable and the stability of the nonlinear system (8.7) could be either stable or unstable.
- If $\cos x_{i1}^* \leq 0$ for all i with strict inequality for at least one i, we can represent Q as in (8.18). Since Φ is a irreducible and nonnegative matrix, it has a positive eigenvalue, r, equal to the spectral radius of Φ, which is between $2k + \min\{-\cos x_{11}^*, \ldots, -\cos x_{N1}^*\}$ and $2k + \max\{-\cos x_{11}^*, \ldots, -\cos x_{N1}^*\}$ ([17], page 537). Therefore, Q has at least one positive eigenvalue. The system is unstable.
- If $\cos x_{i1}^*, i = 1, \ldots, N$, have mixed signs, the system could be either stable or unstable and numerical calculations is necessary to determine the sign of the real parts of the eigenvalues of Q.

$$Q = \begin{bmatrix} \kappa - \cos x_{11}^* & \kappa & 0 & \cdots & & 0 \\ \kappa & -\cos x_{21}^* & \kappa & 0 & & \cdots \\ & \vdots & & & & \\ 0 & \cdots & \kappa & -\cos x_{N-1,1}^* & \kappa \\ 0 & \cdots & 0 & \kappa & \kappa - \cos x_{N1}^* \end{bmatrix}$$

$$+ (-2\kappa)I_N \overset{def}{=} \Phi + (-2\kappa)I_N. \tag{8.18}$$

Remark 8.1   As special cases of Theorem 8.1, the equilibrium points $(2k\pi, 0), k = 0, \pm 1, \ldots$, are asymptotically stable and $((2k + 1)\pi, 0)$ are unstable. The result was first claimed in our early publication [10] without rigorous proof. We extend the result to include all equilibrium points of the open-loop system in this chapter.

8.3.2 Nonlinear Particle Interactions

In the presence of Morse-type nonlinear particle interactions, that is, F_i takes the form (8.3), the equilibrium points of (8.7) are at $(x_{i1}, x_{i2}) = (x_{i1}^*, 0)$ where x_{i1}^* are solutions to

$$- \sin x_{11}^* + \frac{\kappa}{\beta} \left\{ e^{-\beta(x_{21}^* - x_{11}^*)} - e^{-2\beta(x_{21}^* - x_{11}^*)} \right\} = 0,$$

$$- \sin x_{i1}^* + \frac{\kappa}{\beta} \left\{ e^{-\beta(x_{i+1,1}^* - x_{i1}^*)} - e^{-2\beta(x_{i+1,1}^* - x_{i1}^*)} \right\}$$

$$- \frac{\kappa}{\beta} \left\{ e^{-\beta(x_{i1}^* - x_{i-1,1}^*)} - e^{-2\beta(x_{i1}^* - x_{i-1,1}^*)} \right\} = 0,$$

$$i = 2, \dots, N - 1,$$

$$- \sin x_{N1}^*$$

$$- \frac{\kappa}{\beta} \left\{ e^{-\beta(x_{N1}^* - x_{N-1,1}^*)} - e^{-2\beta(x_{N1}^* - x_{N-1,1}^*)} \right\} = 0. \tag{8.19}$$

Let $z_{i1} = x_{i1} - x_{i1}^*, z_{i2} = x_{i2}$, and linearize the system around its equilibrium. After simplification, we get

$$\dot{z}_{i1} = z_{i2}$$

$$\dot{z}_{i2} = - \cos x_{i1}^* z_{i1} - \gamma z_{i2}$$

$$+ \frac{\kappa}{\beta} \left[-e^{-\beta(x_{i+1,1}^* - x_{i1}^*)} + 2e^{-2\beta(x_{i+1,1}^* - x_{i1}^*)} \right] (z_{i+1,1} - z_{i1})$$

$$- \frac{\kappa}{\beta} \left[-e^{-\beta(x_{i1}^* - x_{i-1,1}^*)} + 2e^{-2\beta(x_{i1}^* - x_{i-1,1}^*)} \right] (z_{i1} - z_{i-1,1})$$

$$\stackrel{def}{=} - \cos x_{i1}^* z_{i1} - \gamma z_{i2} + c_{i1}(z_{i+1,1} - z_{i1})$$

$$- c_{i2}(z_{i1} - z_{i-1,1}). \tag{8.20}$$

We can see that (8.20) is in the same form as in (8.9) with different coupling coefficients. We can represent (8.20) as

$$\dot{z} = \overline{A}z + \overline{B}\,\overline{Q}z$$

with the same forms of the matrices $\overline{A}, \overline{B}, \overline{Q}$ as in Section 8.3.1 but different matrix Q shown in (8.21).

$$Q = \begin{bmatrix} Q_{11} & c_{11} & 0 & \cdots & & 0 \\ c_{21} & Q_{22} & c_{22} & 0 & \cdots \\ & & \vdots & & \\ 0 & \cdots & c_{N-1,1} & Q_{N-1,N-1} & c_{N-1,2} \\ 0 & \cdots & & 0 & c_{N2} & Q_{NN} \end{bmatrix}, \tag{8.21}$$

where $Q_{11} = -c_{11} - \cos x_{11}^*$,

$$Q_{22} = -(c_{21} + c_{22} + \cos x_{21}^*),$$

$$Q_{N-1,N-1} = -(c_{N-1,1} + c_{N-1,2} + \cos x_{N-1,1}^*),$$

$$Q_{NN} = -c_{N2} - \cos x_{N1}^*.$$

Following the same procedure as shown in the proof of Theorem 8.1, we conclude that the system (8.7) with nonlinear particle interaction (8.3) is locally asymptotically stable at the equilibrium points $(x_{i1}^*, 0)$ if all of the eigenvalues of the matrix Q defined in (8.21) have negative real parts; it is unstable if any of the eigenvalues of the matrix Q has a positive real part.

In Section 8.4, we formulate our control problem and then discuss the control design in Section 8.5.

8.4 Control Problem Formulation

Control can be applied to the particle array, so that the frictional dynamics of a small array of particles is controlled toward preassigned values of the average sliding velocity. Let the external force, f, in (8.2) be a feedback control, denoted by $u(t)$. Rewrite the system model (8.2) as follows [3]:

$$\ddot{\phi}_i + \gamma \dot{\phi}_i + \sin(\phi_i) = F_i + u(t) \tag{8.22}$$

Due to physical accessibility constraints, the feedback control $u(t)$ is a function of three measurable quantities, v_{target}, $v_{\text{c.m.}}$, and $\phi_{\text{c.m.}}$, where v_{target} is the constant targeted velocity for the center of mass, $v_{\text{c.m.}}$ is the average (center of mass) velocity, that is,

$$v_{\text{c.m.}} = \frac{1}{N} \sum_{i=1}^{N} \dot{\phi}_i, \tag{8.23}$$

and $\phi_{\text{c.m.}}$ is the average (center of mass) position, that is,

$$\phi_{\text{c.m.}} = \frac{1}{N} \sum_{i=1}^{N} \phi_i. \tag{8.24}$$

We define the following tracking control problem:
Design a feasible feedback control law

$$u(t) = u(v_{\text{target}}, v_{\text{c.m.}}, \phi_{\text{c.m.}}), \tag{8.25}$$

such that $v_{\text{c.m.}}$ tends to v_{target}.

In nanoscale friction control, it is sufficient to control the system as a whole. It can be seen that the tracking control problem is a constrained control problem since the accessible variables are average quantities only. Existing results in nonlinear decentralized control (e.g., Refs. [20–23]) cannot be applied due to the inaccessibility of the subsystem' states.

In Section 8.5, we first construct feedback control laws to solve the tracking control problem defined above, and we further analyze the single particle stability in the closed-loop system under the designed average control law.

8.5 Tracking Control Design

To design feedback tracking controllers, we define the following tracking error states:

$$e_{i1} = \phi_i - v_{\text{target}}t, \quad e_{i2} = \dot{\phi}_i - v_{\text{target}}. \tag{8.26}$$

The corresponding error dynamics for a single particle is given as follows:

$$
\begin{aligned}
\dot{e}_{i1} &= e_{i2} \\
\dot{e}_{i2} &= -\sin(e_{i1} + v_{\text{target}}t) - \gamma(e_{i2} + v_{\text{target}}) \\
&\quad + F_i + u(t).
\end{aligned} \tag{8.27}
$$

8.5.1 Tracking Control of the Average System

In this section, we design tracking control to solve the problem defined in Section 8.4, which is to render the average velocity of the system, that is, the velocity of the center of the mass, to converge to a constant targeted value. To this end, we introduce the average error states as

$$e_{1av} = \phi_{\text{c.m.}} - v_{\text{target}}t, \qquad e_{2av} = v_{\text{c.m.}} - v_{\text{target}}, \tag{8.28}$$

where $v_{\text{c.m.}}$ and $\phi_{\text{c.m.}}$ are defined in (8.23) and (8.24), respectively. Then it is obvious that the convergence of $(\phi_{\text{c.m.}}, v_{\text{c.m.}})$ to $(v_{\text{target}}t, v_{\text{target}})$ is equivalent to the convergence of (e_{1av}, e_{2av}) to $(0, 0)$. Therefore, asymptotic stability of the system in the error state space is equivalent to asymptotic tracking of the targeted positions and constant velocity.

The dynamics of (e_{1av}, e_{2av}) can be derived as follows:

$$
\begin{aligned}
\dot{e}_{1av} &= e_{2av} \\
\dot{e}_{2av} &= -\frac{1}{N}\sum_{i=1}^{N}\sin(e_{i1} + v_{\text{target}}t) - \gamma(e_{2av} + v_{\text{target}}). \\
&\quad + u(t)
\end{aligned} \tag{8.29}
$$

Note that the F_i term disappeared in (8.29) because the sum of F_i is zero for Morse-type interactions of the form defined in (8.3).

We construct the following Lyapunov function candidate:

$$W(e_{av}) = \frac{1}{2}e_{1av}^2 + \frac{1}{2}(c_1 e_{1av} + e_{2av})^2, \tag{8.30}$$

where c_1 is a positive design constant, and $e_{av} = [e_{1av} \; e_{2av}]^T$.

Taking the time derivative of W along the dynamics of (8.29), and denoting

$$\xi = c_1 e_{1av} + e_{2av}, \tag{8.31}$$

we have

$$\dot{W}(e_{av}) = -c_1 e_{1av}^2 + \xi \left[e_{1av} + c_1 e_{2av} - \gamma e_{2av} \right.$$
$$\left. - \frac{1}{N} \sum_{i=1}^{N} \sin(e_{i1} + v_{target}t) - \gamma v_{target} + u(t) \right].$$

$$(8.32)$$

Choose

$$u(t) = \gamma v_{target} - e_{1av} - (c_1 - \gamma)e_{2av}$$
$$\quad - (c_1 + c_2)\xi + \sin(v_{target}t)$$
$$= \gamma v_{target} - k_1(\phi_{c.m.} - v_{target}t)$$
$$\quad - k_2(v_{c.m.} - v_{target}) + \sin(v_{target}t),$$

$$(8.33)$$

where c_2 is a positive design constant, $k_1 = 1 + (c_1 + c_2)c_1$, $k_2 = 2c_1 + c_2 - \gamma$, and the term $\sin(v_{target}t)$ is introduced to enforce the equilibrium of the closed-loop system (8.29) to be the origin.

We obtain the following:

$$\dot{W}(e_{av}) = -c_1(e_{1av}^2 + \xi^2) - c_2\xi^2$$
$$\quad + \xi \frac{1}{N} \sum_{i=1}^{N} \left[-\sin(e_{i1} + v_{target}t) + \sin(v_{target}t) \right]$$
$$\leq -c_1(e_{1av}^2 + \xi^2) - c_2\xi^2 + |\xi|$$
$$\quad \cdot \frac{1}{N} \sum_{i=1}^{N} |-\sin(e_{i1} + v_{target}t) + \sin(v_{target}t)|$$
$$\leq -c_1(e_{1av}^2 + \xi^2) - c_2\xi^2 + 2|\xi|.$$

$$(8.34)$$

Since the maximum of the last two terms is $1/c_2$, we have

$$\dot{W}(e_{av}) \leq -c_1(e_{1av}^2 + \xi^2) + \frac{1}{c_2},$$

$$(8.35)$$

which can be used to prove uniform boundedness of the error system (8.29) as shown in the proof of Theorem 8.2.

To achieve asymptotical tracking, that is, to make the error system (8.29) asymptotically stable, the following switching control law can be used:

$$u(t) = \gamma v_{target} - k_1(\phi_{c.m.} - v_{target}t) - k_2 \left(v_{c.m.} - v_{target} \right)$$
$$\quad + \sin(v_{target}t) - 2sgn(\xi),$$

$$(8.36)$$

where $sgn(\xi)$ denotes the signum function, defined as $sgn(\xi) = 1$ for $\xi > 0$, $sgn(\xi) = -1$ for $\xi < 0$, and $sgn(\xi) = 0$ for $\xi = 0$.

Theorem 8.2 presents the stability results of the closed-loop average error system (8.29).

Theorem 8.2   The feedback control laws (8.33) or (8.36) solve the tracking control of the average system defined in Section 8.4. Using (8.33), the tracking error between the velocity of the center of mass and the targeted velocity is uniformly bounded over time $[0, \infty)$. Under the switching control law (8.36), the tracking error goes to zero asymptotically.

Proof: Using the continuous control law (8.33), for the positive definite Lyapunov function W defined in (8.30), we obtained (8.35). Then,

$$\dot{W}(e_{av}) \leq 0, \qquad \forall \|(e_{1av}, \xi)\| \geq \frac{1}{\sqrt{c_1 c_2}}. \tag{8.37}$$

We conclude that the solutions of the closed-loop system (8.29) and (8.33) are globally uniformly bounded.

To calculate the ultimate bound, we note from (8.30) that

$$\frac{1}{2}\lambda_{\min}(P)\|e_{av}\|^2 \leq W(e_{av}) = \frac{1}{2}e_{av}^T P e_{av}$$

$$\leq \frac{1}{2}\lambda_{\max}(P)\|e_{av}\|^2 \tag{8.38}$$

where $e_{av} = [e_{1av} \ e_{2av}]^T$,

$$P = \begin{bmatrix} 1 + c_1^2 & c_1 \\ c_1 & 1 \end{bmatrix},$$

and $\lambda_{\min}(P)$, and $\lambda_{\max}(P)$ denote the minimum and maximum eigenvalues of the matrix P, respectively. From (8.38), we have

$$\|e_{av}\|^2 \leq \frac{2W(e_{av})}{\lambda_{\min}(P)} = \frac{\|(e_{1av}, \xi)\|^2}{\lambda_{\min}(P)}. \tag{8.39}$$

Due to (8.37), we obtain

$$\dot{W}(e_{av}) \leq 0, \qquad \forall \|e_{av}\| \geq \frac{1}{\sqrt{c_1 c_2 \lambda_{\min}(P)}}. \tag{8.40}$$

The ultimate bound of $\|e_{av}\|$ is given by Khalil ([24], Section 4.8) as follows:

$$b = \sqrt{\frac{\lambda_{\max}(P)}{c_1 c_2 \lambda_{\min}^2(P)}}. \tag{8.41}$$

By choosing c_1, c_2 appropriately (with the price of a large control effort), we can have the error states to be arbitrarily close to zero.

Under the switching control law (8.36) (which is the continuous control (8.33) plus a switching term), substituting (8.36) into (8.32), we get

$$\dot{W}(e_{av}) \leq -c_1(e_{1av}^2 + \xi^2) - c_2\xi^2 + 2|\xi| - \xi 2\mathrm{sgn}(\xi)$$

$$\leq -c_1(e_{1av}^2 + \xi^2), \tag{8.42}$$

which is negative definite. Asymptotic stability of the error system follows from Lyapunov theory.   □

It should be noted that the controller proposed in (8.33) renders the velocity of the average system to go to the targeted value, while the individual particles could have different modes of motion. Next, we investigate stability of single particles in the closed-loop system under the average control law (8.33).

8.5.2 Stability of Single Particles in the Closed-Loop System

We assume linear particle interactions in this section. From (8.5) to (8.26), representing F_i using the error states, we have

$$F_i = \kappa \left(e_{i+1,1} - 2e_{i1} + e_{i-1,1} \right), \quad i = 2, \dots, N-1,$$

$$F_1 = \kappa(e_{21} - e_{11}), F_N = \kappa(e_{N-1,1} - e_{N1}). \tag{8.43}$$

For the convenience of presentation, let

$$\bar{k}_1 = \frac{k_1}{N}, \ \bar{k}_2 = \frac{k_2}{N}. \tag{8.44}$$

Substituting the control law defined in (8.33) into (8.27), we have the state-space model of the closed-loop system in the following form:

$$\dot{e}_{i1} = e_{i2}$$

$$\dot{e}_{i2} = -\gamma e_{i2} + F_i - \bar{k}_1 \left(\sum_{j=1}^{N} e_{j1} \right) - \bar{k}_2 \left(\sum_{j=1}^{N} e_{j2} \right)$$

$$+ \left[\sin(v_{\text{target}} t) - \sin(e_{i1} + v_{\text{target}} t) \right]. \tag{8.45}$$

Linearize the system around the equilibrium $e^* = 0$, where $e = [e_{11}, e_{12}, e_{21}, e_{22}, \dots, e_{N1}, e_{N2}]^T$. Since

$$\sin(v_{\text{target}} t) - \sin(e_{i1} + v_{\text{target}} t)$$

$$= -2 \sin \frac{e_{i1}}{2} \cos \frac{e_{i1} + 2v_{\text{target}} t}{2}, \tag{8.46}$$

we obtain the following linearized model:

$$\dot{e} = Ge, \tag{8.47}$$

where

$$G = I_N \otimes \begin{bmatrix} 0 & 1 \\ 0 & -\gamma \end{bmatrix} + Q \otimes \begin{bmatrix} 0 & 0 \\ \kappa & 0 \end{bmatrix} + \Theta \otimes \begin{bmatrix} 0 & 0 \\ -\bar{k}_1 & -\bar{k}_2 \end{bmatrix}$$

$$+ I_N \otimes \begin{bmatrix} 0 & 0 \\ -\cos v_{\text{target}} t & 0 \end{bmatrix}$$

$$= I_N \otimes \begin{bmatrix} 0 & 1 \\ -\cos v_{\text{target}} t & -\gamma \end{bmatrix} + Q \otimes \begin{bmatrix} 0 & 0 \\ \kappa & 0 \end{bmatrix}$$

$$+ \Theta \otimes \begin{bmatrix} 0 & 0 \\ -\bar{k}_1 & -\bar{k}_2 \end{bmatrix} \tag{8.48}$$

where Θ is the N by N matrix of ones, and

$$Q = \begin{bmatrix} -1 & 1 & 0 & \cdots & 0 \\ 1 & -2 & 1 & 0 & \cdots \\ & & \vdots & & \\ 0 & \cdots & 1 & -2 & 1 \\ 0 & \cdots & 0 & 1 & -1 \end{bmatrix}. \tag{8.49}$$

We have Lemma 8.4.

Lemma 8.4 There exists a similarity transformation such that the matrix G in (8.47) can be transformed to a block diagonal one.

Proof of Lemma 8.4: Note that the matrix $(-Q)$ is a real symmetric matrix with zero row sum, and it is irreducible. From Lemmas 8.1 to 8.3, $(-Q)$ has eigenvalues

$$\mu_1 \geq \mu_2 \geq \cdots \geq \mu_{N-1} > \mu_N = 0. \tag{8.50}$$

It is always possible to choose the eigenvectors to be real, normalized, and mutually orthogonal. Denote the eigenvectors corresponding to each of the eigenvalues:

$$v_k = [v_{1k}, v_{2k}, \ldots, v_{Nk}], k = 1, 2, \ldots, N-1;$$
$$v_N. \tag{8.51}$$

Then $V = [v_1 \; v_2 \; \cdots \; v_N]$ is an orthogonal matrix, that is, $VV^T = V^TV = I$, implying $V^T = V^{-1}$, and

$$\sum_{k=1}^{N} v_{ki}v_{kj} = \sum_{k=1}^{N} v_{ik}v_{jk} = \delta_{ij}, \tag{8.52}$$

where $\delta_{ij} = 1$ for $i = j$ and $\delta_{ij} = 0$ for $i \neq j$. Because of $-V^TQV = \text{diag}(\mu_1, \mu_2, \ldots, \mu_N)$, we further have

$$(-Q)_{ij} = \sum_{k=1}^{N} \mu_k v_{ik}v_{jk}. \tag{8.53}$$

Because the eigenvectors $v_k, k = 1, 2, \ldots, N-1$, are orthogonal to v_N, the following property holds:

$$\sum_{j=1}^{N} v_{jk} = 0, k = 1, 2, \ldots, N-1,$$
$$v_N = \frac{1}{\sqrt{N}}[1 \; 1 \; \cdots \; 1]^T. \tag{8.54}$$

Therefore, we have

$$V^{-1}QV = -D_Q, \tag{8.55}$$

where D_Q is a diagonal matrix with the diagonal entry $\mu_i, i = 1, 2, \ldots, N$.

Due to property (8.54), the matrix V transforms the all 1's matrix Θ to a diagonal one as well.

$$
\begin{aligned}
V^{-1}\Theta V &= \left[(V^{-1}\Theta V)_{ik} \right] \\
&= \left[\left(\sum_{j=1}^{N} v_{ji} \right) \left(\sum_{j=1}^{N} v_{jk} \right) \right], \\
&= D_\Theta
\end{aligned}
\tag{8.56}
$$

where D_Θ is a diagonal matrix with diagonal entry $(D_\Theta)_{ii} = 0, i = 1, 2, \ldots,$ $N - 1$, and $(D_\Theta)_{NN} = N$. Choose the transformation matrix as follows:

$$
T = V \otimes I_2.
\tag{8.57}
$$

We have

$$
\begin{aligned}
T^{-1}GT &= \left(V \otimes I_2 \right)^{-1} \left(I_N \otimes \begin{bmatrix} 0 & 1 \\ -\cos v_{\text{target}}t & -\gamma \end{bmatrix} \right. \\
&\quad \left. + Q \otimes \begin{bmatrix} 0 & 0 \\ \kappa & 0 \end{bmatrix} + \Theta \otimes \begin{bmatrix} 0 & 0 \\ -\bar{k}_1 & -\bar{k}_2 \end{bmatrix} \right) \left(V \otimes I_2 \right) \\
&= I_N \otimes \begin{bmatrix} 0 & 1 \\ -\cos v_{\text{target}}t & -\gamma \end{bmatrix} - D_Q \otimes \begin{bmatrix} 0 & 0 \\ \kappa & 0 \end{bmatrix} \\
&\quad + D_\Theta \otimes \begin{bmatrix} 0 & 0 \\ -\bar{k}_1 & -\bar{k}_2 \end{bmatrix} \\
&= \operatorname{diag}\{C_i\},
\end{aligned}
\tag{8.58}
$$

where

$$
C_i = \begin{cases} \begin{bmatrix} 0 & 1 \\ -\cos v_{\text{target}}t - \mu_i \kappa & -\gamma \end{bmatrix}, & i = 1, 2, \ldots, N-1, \\ \begin{bmatrix} 0 & 1 \\ -\cos v_{\text{target}}t - k_1 & -k_2 - \gamma \end{bmatrix}, & i = N. \end{cases}
\tag{8.59}
$$

This completes the proof of the lemma. □

We are now in the position to state the main theorem of this section.

Theorem 8.3 For system parameters κ and γ that satisfy

$$
\kappa > \frac{1}{\mu_{N-1}}, \quad \gamma > \frac{v_{\text{target}}}{2(\mu_{N-1}\kappa - 1)},
\tag{8.60}
$$

where μ_{N-1} is the second smallest eigenvalue of the matrix $(-Q)$, choose the control parameters

$$
k_1 > 1, \quad k_2 > max \left\{ \frac{v_{\text{target}}}{2k_1} - \gamma, 0 \right\},
\tag{8.61}
$$

then the error system for individual particles (8.45) is locally asymptotically stable.

Proof of Theorem 8.3: We use the classic Lyapunov theory to prove the local stability of the error system (8.45).

From Lemma 8.4, under similarity transformation $z = T^{-1}e$, system (8.47) is transferred to the following one:

$$\dot{z} = \text{diag}\{C_i\}z. \tag{8.62}$$

Here, C_i is represented in (8.59).

Define the following Lyapunov function candidate:

$$W(t,z) = \sum_{i=1}^{N} \left\{ \frac{\varepsilon_i}{2}z_{i1}^2 + \frac{1}{2}(\lambda_i z_{i1} + z_{i2})^2 \right\}$$

$$+ \sum_{i=1}^{N} \left\{ \frac{1}{2}[1 + \cos(v_{\text{target}}t)]z_{i1}^2 \right\}, \tag{8.63}$$

where ε_i, and $\lambda_i, i = 1, \ldots, N$, are design parameters.

We can see that

$$W_1(z) \le W(t,z) \le W_2(z), \tag{8.64}$$

where $W_1(z)$ and $W_2(z)$ are both positive definite.

$$W_1(z) = \sum_{i=1}^{N} \left\{ \frac{\varepsilon_i}{2}z_{i1}^2 + \frac{1}{2}(\lambda_i z_{i1} + z_{i2})^2 \right\}$$

$$W_2(z) = \sum_{i=1}^{N} \left\{ \left(1 + \frac{\varepsilon_i}{2}\right)z_{i1}^2 + \frac{1}{2}(\lambda_i z_{i1} + z_{i2})^2 \right\}. \tag{8.65}$$

Take the time derivative of $W(t,z)$ along the system dynamics (8.62). We have

$$\dot{W}(t,z) = \sum_{i=1}^{N-1} \left[\varepsilon_i + \lambda_i(\lambda_i - \gamma) - \mu_i\kappa + 1\right]z_{i1}z_{i2}$$

$$+ \sum_{i=1}^{N-1} \left[-\frac{1}{2}v_{\text{target}}\sin(v_{\text{target}}t) - \lambda_i\cos(v_{\text{target}}t) - \lambda_i\mu_i\kappa\right]z_{i1}^2$$

$$+ \sum_{i=1}^{N-1}(\lambda_i - \gamma)z_{i2}^2$$

$$+ \frac{1}{2}[\varepsilon_N + \lambda_N^2 - \lambda_N(k_2 + \gamma) - k_1 + 1]z_{N1}z_{N2}$$

$$- \lambda_N k_1 z_{N1}^2 - (k_2 + \gamma - \lambda_N)z_{N2}^2$$

$$- \frac{1}{2}\sin(v_{\text{target}}t)v_{\text{target}}z_{N1}^2. \tag{8.66}$$

Under the condition on κ,

$$\kappa > \frac{1}{\min_{i \leq N-1}(\mu_i)} = \frac{1}{\mu_{N-1}}, \tag{8.67}$$

we have $\mu_i \kappa > 1$. Choose the design and control parameters,

$$\lambda_i < \gamma, \tag{8.68}$$

$$\varepsilon_i = \lambda_i(\gamma - \lambda_i) + \mu_i \kappa - 1 > 0, \quad i = 1, 2, \ldots, N-1 \tag{8.69}$$

$$\lambda_N < \gamma + k_2, \tag{8.70}$$

$$\varepsilon_N = \lambda_N(k_2 + \gamma - \lambda_N) + k_1 - 1 > 0, \tag{8.71}$$

$$k_1 > 1, \tag{8.72}$$

so that the cross terms are zero. Bounding the sinusoidal terms, we obtain

$$\dot{W}(t, z) \leq - \sum_{i=1}^{N-1} \left(\lambda_i \mu_i \kappa - \frac{1}{2} v_{\text{target}} - \lambda_i \right) z_{i1}^2$$

$$- \sum_{i=1}^{N-1} (\gamma - \lambda_i) z_{i2}^2 - (\lambda_N k_1 - \frac{1}{2} v_{\text{target}}) z_{N1}^2$$

$$- (k_2 + \gamma - \lambda_N) z_{N2}^2. \tag{8.73}$$

Because of the following condition on γ,

$$\gamma > \frac{v_{\text{target}}}{2 \min_{i \leq N-1}(\mu_i \kappa - 1)} = \frac{v_{\text{target}}}{2(\mu_{N-1} \kappa - 1)}, \tag{8.74}$$

there exists $\lambda_i < \gamma, i = 1, \ldots, N-1$, such that

$$\lambda_i \mu_i \kappa - \frac{1}{2} v_{\text{target}} - \lambda_i > 0. \tag{8.75}$$

To get $\lambda_N k_1 - \frac{1}{2} v_{\text{target}} > 0$, we need $\lambda_N > v_{\text{target}}/(2k_1)$. Combing with (8.70), we need to choose the following control parameter so that λ_N exists:

$$k_2 > \max \left\{ \frac{v_{\text{target}}}{2k_1} - \gamma, 0 \right\}. \tag{8.76}$$

Therefore, we have

$$\dot{W}(t, z) \leq -W_3(z), \tag{8.77}$$

where $W_3(z)$ is positive definite.

The asymptotical stability of (8.62) follows directly from the Lyapunov stability theory because of (8.64) and (8.77) ([24], Chapter 4). Due to the similarity transformation $z = T^{-1}e$, (8.47) is asymptotically stable. Therefore, the non-linear error system (8.45) is locally asymptotically stable. □

Remark 8.2   Since the application of Lyapunov direct method provides sufficient conditions only, the obtained condition on γ and κ is sufficient and may be conservative.

8.6 Simulation Results

We have performed extensive numerical simulations on arrays of different sizes ($3 \leq N \leq 256$). The system parameters used are $\gamma = 0.1$, $\kappa = 0.26$ [3, 25]. Random initial conditions are used in the simulations.

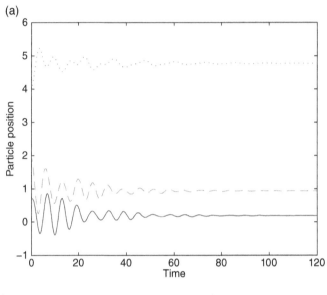

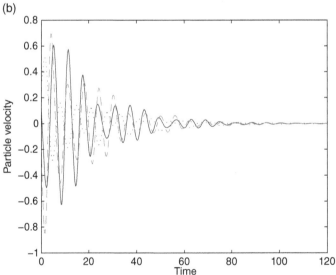

Figure 8.2 Local stability of the equilibrium points $(\phi_1, \dot{\phi}_1, \phi_2, \dot{\phi}_2, \phi_3, \dot{\phi}_3) =$ (0.1941, 0, 0.9360, 0, 4.7747, 0) in the presence of linear particle interactions: (a) particle positions and (b) particle velocities.

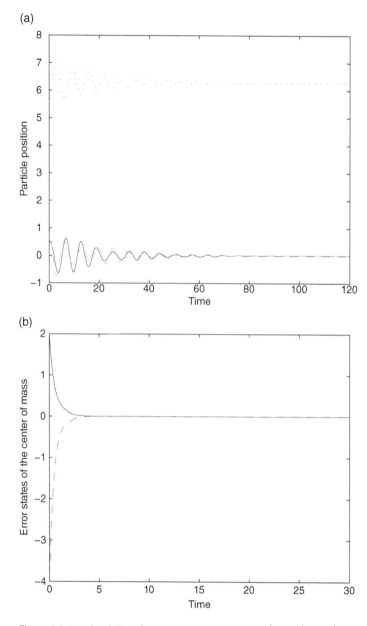

Figure 8.3 Local stability of the equilibrium points $(\phi_1, \dot{\phi}_1, \phi_2, \dot{\phi}_2, \phi_3, \dot{\phi}_3) =$ (0.0001, 0, 0.0004, 0, 6.2827, 0) in the presence of nonlinear particle interactions: (a) particle positions and (b) particle velocities.

First, we verify the stability of the open-loop frictional dynamics (8.7). In the presence of linear particle interactions, for the system parameters

$$N = 3, \kappa = 0.26, \gamma = 0.1,$$

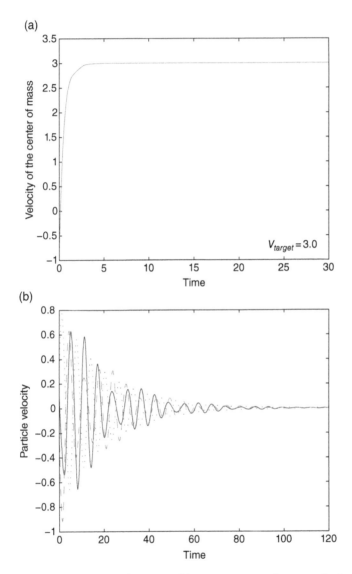

Figure 8.4 Tracking performance of the average system for targeted value $v_{target} = 3$: (a) the time history of the velocity of the center of the mass, (b) the time history of the error states of the center of the mass with the solid line denoting e_{1av} and the dashed line denoting e_{2av}, and (c) the control history.

(c)

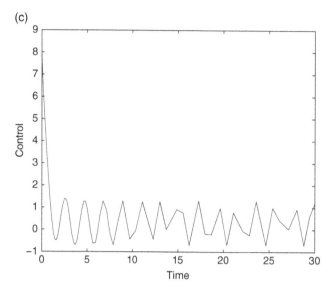

Figure 8.4 (*Continued*).

we have one set of the equilibrium points (by solving (8.8)) is at

$$(\phi_1, \dot{\phi}_1, \phi_2, \dot{\phi}_2, \phi_3, \dot{\phi}_3) = (0.1941, 0, 0.9360, 0, 4.7747, 0).$$

We check that $\phi_i \in (\frac{2k+1}{2}\pi, \frac{2k+3}{2}\pi), k = -1, 1, i = 1, 2, 3$ and the matrix Q defined in (8.13) is negative definite. According to Theorem 8.1, the open-loop system is locally asymptotically stable at these points. This is verified by Figure 8.2.

In the presence of nonlinear particle interactions, with the same system parameters κ, γ and $\beta = 1$, we have one set of the equilibrium points (by solving (8.19)) is at

$$(\phi_1, \dot{\phi}_1, \phi_2, \dot{\phi}_2, \phi_3, \dot{\phi}_3) = (0.0001, 0, 0.0004, 0, 6.2827, 0).$$

We checked that the matrix Q defined in (8.21) is negative definite, so the system is locally asymptotically stable at these points. This is verified by Figure 8.3.

Figures 8.4 and 8.5 demonstrate the tracking performances of the average system using the control law (8.33) with different initial conditions and for two different targeted values $v_{\text{target}} = 3, 1.5$, respectively. In both Figures 8.4 and 8.5, (a) shows the time history of the velocity of the center of the mass, that is, $v_{\text{c.m.}}$, (b) shows the error states of the center of the mass, that is, e_{1av} and e_{2av}, and (c) shows the control history.

As pointed out in Section 8.5.1, the individual particle could be in different modes of motion, while the average system tracks a targeted trajectory. Figure 8.6 shows the oscillating dynamics of individual particles of the closed-loop system tracking the targeted average velocity $v_{\text{target}} = 1.5$ whose average tracking

(a)

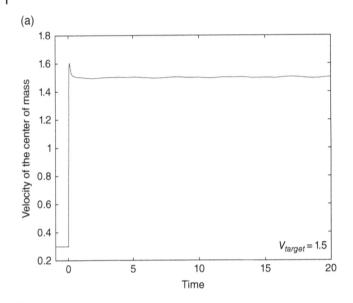

(b)

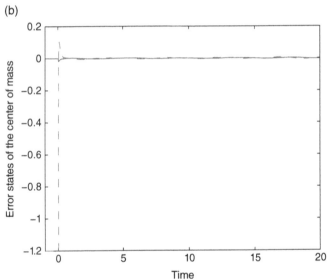

Figure 8.5 Tracking performance of the average system for targeted value $v_{target} = 1.5$: (a) the time history of the velocity of the center of the mass, (b) the time history of the error states of the center of the mass with the solid line denoting e_{1av} and the dashed line denoting e_{2av}, and (c) the control history.

(c)

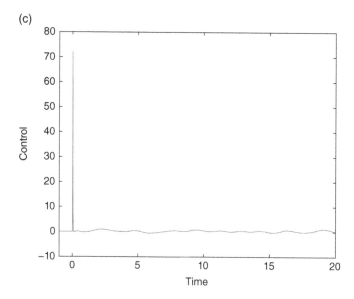

Figure 8.5 (*Continued*).

performance was shown in Figure 8.5. It can be seen that individual particles are oscillating, while the velocity of the average system tracks the targeted value.

In comparison, we show the stability of the error states of individual particles under the control law (8.33) in Figure 8.7. The system parameters used are $\gamma = 1.6, \kappa = 1.5$ for a three-particle interconnected system. The control parameters are chosen to be $k_1 = 1.2, k_2 = 0.4$. The targeted velocity is $v_{\text{target}} = 1.5$. Figure 8.7 shows that the tracking error for each individual particle tends to zero which indicates that the velocity of each particle in the interconnected system tracks the targeted value. This verifies the result in Theorem 8.3.

8.7 Conclusion

We studied the stability and the control problem for a 1D particle array sliding on a surface subject to friction. The well-known FK model is used to describe the dynamics, which represents a nonlinear interconnected system. A control problem is formulated and the control objective is for the average system to reach a designated targeted velocity using physically accessible variables, that is, the average quantity of the system. Local stability of the unforced system at different equilibrium points is revealed first in the presence of linear and nonlinear particle interconnections, respectively. A global feedback control law is then constructed to achieve the control objective by utilizing the Lyapunov theory–based method. Simulation results are shown to illustrate satisfactory performances. The results of the chapter can be applied to other physical systems whose dynamics can be described by the FK model.

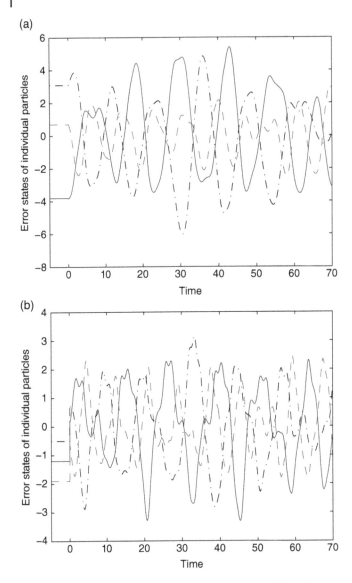

Figure 8.6 Particles dynamics of the average system tracking the targeted average velocity $v_{target} = 1.5$: (a) the phase variables of individual particles and (b) the velocity variables of individual particles.

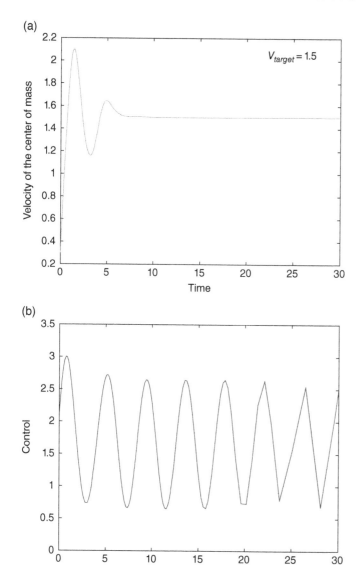

Figure 8.7 Stability of individual particles in the closed-loop tracking system for targeted value $v_{target} = 1.5$: (a) the time history of the velocity of the center of the mass, (b) the control history, (c) the error phase variables of individual particles and (d) the error velocity variables of individual particles.

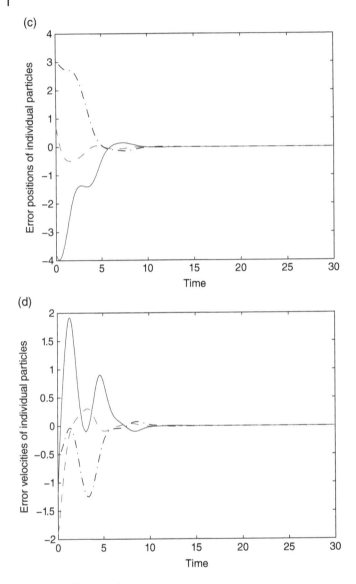

Figure 8.7 (*Continued*).

Notes

The results in this chapter were first published in Ref. [26]. Further study of atomic-scale friction control using friction force microscope was reported in Ref. [27].

Acknowledgment is given to ©2015 Elsevier. Reprinted, with permission, from Guo, Y. and Qu, Z. (2008) Control of frictional dynamics of a one-dimensional particle array. Automatica, 44, 2560–2569.

References

1 Persson, B.N.J. (2000) *Sliding Friction*, 2nd edn., Springer, Berlin/Heidelberg.
2 Urbakh, M., Klafter, J., Gourdon, D., and Israelachvili, J. (2004) The nonlinear nature of friction. *Nature*, **430**, 525–528.
3 Braiman, Y., Barhen, J., and Protopopescu, V. (2003) Control of friction at the nanoscale. *Physical Review Letters*, **90** (9), 094301.
4 Cochard, A., Bureau, L., and Baumberger, T. (2003) Stabilization of frictional sliding by normal load modulation: A bifurcation analysis. *Transactions of the ASME*, **70**, 220–226.
5 Gao, J., Luedtke, W., and Landman, U. (1998) Friction control in thin-film lubrication. *Journal of Physical Chemistry B*, **102**, 5033–5037.
6 Heuberger, M., Drummond, C., and Israelachvili, J.N. (1998) Coupling of normal and transverse motions during frictional sliding. *Journal of Physical Chemistry B*, **102**, 5038–5041.
7 Rozman, M.G., Urbakh, M., and Klafter, J. (1998) Controlling chaotic friction. *Physical Review E*, **57**, 7340–7443.
8 Zaloj, V., Urbakh, M., and Klafter, J. (1999) Modifying friction by manipulating normal response to lateral motion. *Phyical Review Letters*, **82**, 4823–4826.
9 Protopopescu, V. and Barhen, J. (2004) Non-Lipschitzian control algorithms for extended mechanical systems. *Chaos*, **14** (2), 400–407.
10 Guo, Y., Qu, Z., and Zhang, Z. (2006) Lyapunov stability and precise control of the frictional dynamics of a one-dimensional nanoarray. *Physical Review B*, **73** (9), 094 118.
11 Braiman, Y., Family, F., and Hentschel, H.G.E. (1997) Nonlinear friction in the periodic stick-slip motion of coupled oscillators. *Physical Review B*, **55** (5), 5491–5504.
12 Braun, O.M. and Kivshar, Y.S. (2004) *The Frenkel-Kontorova Model*, Springer, Berlin/Heidelberg.
13 Chou, C.I., Ho, C.L., Hu, B., and Lee, H. (1998) Morse-type Frenkel-Kontorova model. *Physical Review E*, **57** (3), 2747–2756.
14 Gazi, V. and Passino, K.M. (2002) A class of attraction/repulsion functions for stable swarm aggregations, in *Proceedings of the IEEE Conference on Decision and Control*, Las Vegas, Nevada, pp. 2842–2847.
15 Gazi, V. and Passino, K.M. (2003) Stability analysis of swarms. *IEEE Transactions on Automatic Control*, **48** (4), 692–696.
16 Godsil, C. and Royle, G. (2001) *Algebraic Graph Theory*, Springer, New York.

17 Lancaster, P. and Tismenetsky, M. (1985) *The Theory of Matrices with Applications*, 2nd edn., Academic Press, San Diego, CA.

18 Wu, C.W. (2002) *Synchronization in Coupled Chaotic Circuits and Systems*, World Scientific, Singapore.

19 Slotine, J.J.E. and Li, W. (1991) *Applied Nonlinear Control*, Prentics Hall, Englewood Cliffs, NJ.

20 Guo, Y., Jiang, Z.P., and Hill, D.J. (1999) Decentralized robust disturbance attenuation for a class of large-scale nonlinear systems. *Systems and Control Letters*, **37**, 71–85.

21 Ioannou, P. (1986) Decentralized adaptive control of interconnected systems. *IEEE Transactions on Automatic Control*, **31** (4), 291–298.

22 Jiang, Z.P. (2002) Decentralized disturbance attenuating output-feedback trackers for large-scale nonlinear systems. *Automatica*, **38**, 1407–1415.

23 Tezcan, I.E. and Basar, T. (1999) Disturbance attenuating adaptive controllers for parametric strict feedback nonlinear systems with output measurements. *ASME Journal of Dynamic Systems, Measurement and Control*, **121**, 48–57.

24 Khalil, H. (2002) *Nonlinear Systems*, 3rd edn., Prentice Hall, Upper Saddle River, NJ.

25 Guo, Y. and Qu, Z. (2005) Stabilization and tracking control of friction dynamics of a one-dimensional nanoarray, in *Proceedings of the American Control Conference*, Portland, OR, pp. 2487–2492.

26 Guo, Y. and Qu, Z. (2008) Control of frictional dynamics of a one-dimensional particle array. *Automatica*, **44**, 2560–2569.

27 Guo, Y., Wang, Z., Qu, Z., and Braiman, Y. (2011) Atomic-scale friction control by vibration using friction force microscope. *IFAC Control Engineering Practice*, **19** (11), 1387–1397.

9

Synchronizing Coupled Semiconductor Lasers

9.1 Introduction

Laser diodes (LDs) are compact optical devices that impact large variety of applications including optical communications and optical storage. Despite the advancement in LD production, the output power from single-mode LD remains quite limited. Coherent beam combining provides a viable path to increase coherent emission power from many small lasers. As a consequence of coherent beam combination, in-phase locking of LDs can be realized resulting in a constructive interference along the optical axis. Achieving phase synchronization of LDs poses a very intriguing challenge. Synchronization of semiconductor LD array (LDA) comprising single-mode LDs has been investigated theoretically predominantly employing the nearest-neighbor and global coupling between the lasers [1–4]. General coupling topology has not been explored.

The dynamics of each element of the semiconductor laser array is commonly described by the Lang and Kobayashi equations [5]. In the case of global coupling, the process of synchronization shows analogy to the process found in the Kuramoto model [6]. It is unknown whether this result holds for general coupling topologies of laser arrays. While Kuramoto model describes synchronization behaviors of coupled *phase* oscillators, *coupled laser arrays* have a highly nonlinear model and represent a complex system with both technological [7] and theoretical [8–12] importance. From a theoretical perspective, laser arrays provide a prime example of coupled limit-cycle oscillators, which connects to explorations of pattern formation and many other topics throughout physics, chemistry, biology, and engineering [6, 13].

In this chapter, we investigate general coupling topologies for coupled semiconductor laser arrays and characterize synchronization conditions. We use graph Laplacian tools inspired by recent cooperative control advances. Examining the dynamic model of coupled semiconductor lasers described by the Lang and Kobayashi equations [5], we first analyze decoupled laser dynamics and reveal local stability properties, and then study coupled laser arrays under general coupling topologies. We characterize synchronization

Distributed Cooperative Control: Emerging Applications, First Edition. Yi Guo.
© 2017 John Wiley & Sons, Inc. Published 2017 by John Wiley & Sons, Inc.
Companion website: www.wiley.com/go/Guo/DistributedCooperativeControl

conditions using graph and systems theory. The results not only advance current synchronization methods for semiconductor laser arrays but also provide a real-world example of coupled high-dimensional nonlinear systems for cooperative control study.

Organization: The rest of the chapter is organized as follows. Section 9.2 presents the model of semiconductor lasers. Then stability analysis of a single laser system is conducted in Section 9.3, which is followed by synchronization analysis of coupled semiconductor laser arrays in Section 9.4. Simulations are performed in Section 9.5. Section 9.6 concludes the chapter.

9.2 The Model of Coupled Semiconductor Lasers

We consider a coupled semiconductor laser system where each laser is subject to the optical feedback reflected from a mirror. Figure 9.1 shows a globally optical coupled semiconductor array as appeared in Ref. [4]. We study the following dynamic equations for n linearly coupled semiconductor lasers [5, 14–16]:

$$\dot{E}_k = \frac{1+i\alpha}{2} \left[\frac{g(N_k - N_0)}{1+s|E_k|^2} - \gamma \right] E_k + i\omega E_k$$

$$+ \frac{K}{m_k} \sum_{j \in \mathcal{N}(k) \cup \{k\}} E_j$$

$$\dot{N}_k = J - \gamma_n N_k - \frac{g(N_k - N_0)}{1+s|E_k|^2} |E_k|^2, \tag{9.1}$$

where $k = 1, \ldots, n$, $\mathcal{N}(k)$ denotes the neighbor set of the kth laser, $m_k = |\mathcal{N}(k)| + 1$ with $|\mathcal{N}(k)|$ denoting the number of neighbors of the kth laser, E_k is the complex electric field of laser k, and i is the imaginary unit, that

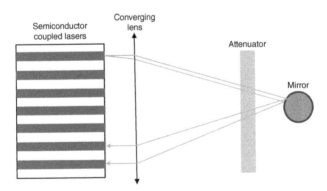

Figure 9.1 Schematic representation of a semiconductor coupled laser array with global optical coupling between the lasers [4].

is, $\sqrt{-1} = i$. N_k is the carrier number for laser k, ω is the oscillating frequency, $J, K > 0$ are constant pump current and coupling strength, respectively, both of which are assumed to be identical for all lasers. $\alpha, N_0, s, \gamma, \gamma_n$, and g are system parameters and are all positive, whose physical meaning will be given later in Section 9.5.

Remark 9.1   As a special case of all-to-all coupling, the last term in the first equation of (9.1) becomes $\frac{K}{n} \sum_{j=1}^{n} E_j$, which is consistent with the global coupling model in the literature (*e.g.*, Refs. [4] and [17]).

Remark 9.2   Figure 9.1 is an example configuration of globally coupled semiconductor arrays. For different coupling topologies, its physical implementation may be different. To focus on stability analysis utilizing graph Laplacian tools, we do not consider time delays in our equation, which is practically unavoidable in some real laser systems (i.e., external cavity systems [18]). Also, experimental configurations of various coupling topologies are out of the scope of the chapter.

We have the following assumption on the coupling topology.

Assumption 9.1   Assume each laser has the same number of neighboring connections, and the coupling topology between lasers can be described by a general undirected connected graph with a Laplacian matrix L.

This assumption indicates $|\mathcal{N}(k)| = |\mathcal{N}(j)|$ for any j, k, and $m_i = m_j = m$. Recall that the in-degree (or equivalently out-degree for undirected graphs) is defined to be the number of connections of each node, so that Assumption 9.1 indicates that each node of the laser network has the same in-degree. We define the Laplacian matrix as $L = [L_{ij}]$ with $L_{ij} = -\frac{1}{m}$ for $j \neq i, j \in \mathcal{N}(i)$, $L_{ij} = 0$ for $j \neq i, j \notin \mathcal{N}(i)$, and $L_{ii} = \frac{m-1}{m}$.

Inspired by the consensus-based work, we rewrite the dynamic equation (9.1) in the following form with a coupling term in the form of relative electrical field:

$$\dot{E}_k = \left[\frac{1 + i\alpha}{2} \left(\frac{g(N_k - N_0)}{1 + s|E_k|^2} - \gamma \right) + K \right] E_k + i\omega E_k$$
$$+ \frac{K}{m} \sum_{j \in \mathcal{N}(k)} (E_j - E_k)$$
$$\dot{N}_k = J - \gamma_n N_k - \frac{g(N_k - N_0)}{1 + s|E_k|^2} |E_k|^2 \qquad (9.2)$$

To facilitate analysis, we represent the electrical field of each laser as $E_k = r_k e^{i\theta_k} = r_k \cos\theta_k + i r_k \sin\theta_k$ in polar coordinates. We obtain the coupled laser dynamics as

$$\dot{r}_k = \frac{1}{2}\left(\frac{g(N_k - N_0)}{1 + sr_k^2} - \gamma + 2K\right)r_k$$

$$+\frac{K}{m}\sum_{j\in\mathcal{N}(k)}(r_j\cos(\theta_j - \theta_k) - r_k)$$

$$\dot{\theta}_k = \frac{\alpha}{2}\left(\frac{g(N_k - N_0)}{1 + sr_k^2} - \gamma\right) + \omega$$

$$+\frac{K}{m}\sum_{j\in\mathcal{N}(k)}\frac{r_j}{r_k}\sin(\theta_j - \theta_k)$$

$$\dot{N}_k = J - \gamma_n N_k - \frac{g(N_k - N_0)}{1 + sr_k^2}r_k^2, \tag{9.3}$$

where $k = 1, \ldots, n$.

Assume identical frequency ω for each laser, we are interested in the synchronization behaviors. We define laser synchronization as the trajectories of all the lasers approaching each other.

Definition 9.1   The system (9.1) is said to synchronize if

$$E_k(t) \to E_j(t), \quad \forall k, j = 1, \ldots, n, \text{ as } t \to \infty. \tag{9.4}$$

The problem of interest is to characterize the conditions so that the system (9.1) synchronizes.

Note that the given equations describe a three-dimensional dynamic system of coupled lasers with a given oscillating frequency ω. Before investigating dynamics of the coupled laser system, in Section 9.3 we start to investigate the stability property of a decoupled laser, that is, the dynamics without the coupling terms in Equation (9.3).

9.3   Stability Properties of Decoupled Semiconductor Laser

In this section, we analyze local stability of a decoupled laser, which is described by the Equation (9.3) without the coupling terms. Note that this is also the dynamics of coupled lasers when they're fully synchronized, that is when $E_1 = E_2 = \cdots = E_n$. We rewrite the decoupled laser dynamics as follows:

$$\dot{r} = f_r(r, N) = \frac{1}{2}\left(\frac{g(N - N_0)}{1 + sr^2} - \gamma + 2K\right)r$$

$$\dot{\theta} = f_\theta(r, N) = \frac{\alpha}{2}\left(\frac{g(N - N_0)}{1 + sr^2} - \gamma\right) + \omega$$

$$\dot{N} = f_N(r, N) = J - \gamma_n N - \frac{g(N - N_0)}{1 + sr^2}r^2. \tag{9.5}$$

We note there exists an equilibrium solution to

$$\dot{r}^* = \frac{1}{2}\left(\frac{g(N^* - N_0)}{1 + sr^{*2}} - \gamma + 2K\right)r^* = 0$$

$$\dot{\theta}^* = -\alpha K + \omega$$

$$\dot{N}^* = J - \gamma_n N^* - \frac{g(N^* - N_0)}{1 + sr^{*2}}r^{*2} = 0 \tag{9.6}$$

in terms of electrical field with constant amplitude r^* and a constant angular speed ω. The corresponding trajectory of the system is on a circle with $(r, \theta, N) = (r^*, \theta^*, N^*)$, where (r^*, θ^*, N^*) is the nontrivial solution of the given equations, which can be computed as:

$$r^* = \sqrt{\frac{gJ - g\gamma_n N_0 - (\gamma - 2K)\gamma_n}{(\gamma - 2K)(\gamma_n s + g)}} \tag{9.7a}$$

$$\theta^* = (-\alpha K + \omega)t + \theta_0 \tag{9.7b}$$

$$N^* = \frac{sJ + \gamma - 2K + gN_0}{g + s\gamma_n}, \tag{9.7c}$$

where θ_0 is the initial phase angle.

This solution exists only when the pump current J is larger than a threshold Jth given by

$$J\text{th} = \gamma_n(N_0 + (\gamma - 2K)/g). \tag{9.8}$$

To ensure the pump current is positive and the solution to (9.7a) exists, the coupling strength K must satisfy $K < \gamma/2$.

Since in the dynamic equation (9.5), the phase variable θ has no influence to the amplitude variable r and the carrier number variable N, we first analyze the stability of the (r, N) system, and then consider the behavior of the θ dynamics. For this purpose, we linearize the decoupled laser system (9.5) around its equilibrium point $(r, N) = (r^*, N^*)$ to get the linearized form,

$$\dot{r} = \frac{-s\gamma'r^{*2}}{1 + sr^{*2}}(r - r^*) + \frac{gr^*}{2(1 + sr^{*2})}(N - N^*)$$

$$\dot{N} = -\frac{2\gamma'r^*}{1 + sr^{*2}}(r - r^*) + \left(-\gamma_n - \frac{gr^{*2}}{1 + sr^{*2}}\right)(N - N^*),$$

$$\tag{9.9}$$

where $\gamma' = \gamma - 2K$.

Define new states as $\Delta r = r - r^*$, $\Delta N = N - N^*$, the linearized system is rewritten as

$$\frac{d}{dt}\begin{bmatrix} \Delta r \\ \Delta N \end{bmatrix} = A\begin{bmatrix} \Delta r \\ \Delta N \end{bmatrix}, \tag{9.10}$$

where

$$A = \begin{bmatrix} \dfrac{-s\gamma'r^{*2}}{1+sr^{*2}} & \dfrac{gr^*}{2(1+sr^{*2})} \\ \dfrac{2\gamma'r^*}{1+sr^{*2}} & -\gamma_n - \dfrac{gr^{*2}}{1+sr^{*2}} \end{bmatrix}$$

$$\overset{\text{def}}{=} \begin{bmatrix} A_{11} & A_{12} \\ A_{21} & A_{22} \end{bmatrix}. \tag{9.11}$$

We can see that the local stability of the system (9.10) depends on the eigenvalues of the matrix A. Theorem 9.1 provides stability results of the decoupled semiconductor laser system (9.5).

Theorem 9.1 For the system parameter $0 < K < \gamma/2$, the decoupled semiconductor laser system (9.5) locally converges to the limit cycle $(r, N) = (r^*, N^*)$ with the angular velocity $\dot{\theta} = (-\alpha K + \omega)$.

Proof: We first analyze the linearized system (9.10) by solving the equation $|\lambda I - A| = 0$ to find the eigenvalues λ of matrix A. The trajectory of (9.10) is asymptotically stable if the two eigenvalues of A satisfy $Re\{\lambda_1\}, Re\{\lambda_2\} < 0$.

Since λ_1, λ_2 are the roots of the quadratic equation

$$\lambda^2 + \left(\gamma_n + \frac{(g + s\gamma')r^{*2}}{1 + sr^{*2}} \right) \lambda + \frac{(g + \gamma_n s)\gamma'r^{*2}}{1 + sr^{*2}} = 0 \tag{9.12}$$

denoting $a = 1$, b and c as the coefficients of the first-order term and the constant, their values can be computed by $\lambda_1, \lambda_2 = (-b \pm \sqrt{\Delta})/2a$. Whether they are two distinct real roots or two distinct complex roots depends on the sign of the discriminant $\Delta = b^2 - 4ac$ of Equation (9.12). If $b > 0$, $\Delta \geq 0$, the roots are both real. They are negative numbers if and only if $-b + \sqrt{\Delta} < 0$, which implies $b^2 > b^2 - 4ac$. With $a = 1$, we have $0 < c \leq b^2/4$. If $\Delta < 0$, the roots are two distinct complex roots with negative real parts. In this case, we have $c > b^2/4a = b^2/4$. If $b \leq 0$, for any value of Δ, the roots can not both have negative real parts. Following the given statement, the condition for $Re\{\lambda_1\}, Re\{\lambda_2\} < 0$ is $b, c > 0$. For arbitrary positive system parameters $g, s, \alpha, \gamma, \gamma_n, N_0 > 0, J > T$th, together with $r^{*2} > 0$, the condition

$$c = \frac{(g + \gamma_n s)\gamma'r^{*2}}{1 + sr^{*2}} > 0 \tag{9.13}$$

is satisfied if and only if $\gamma' = \gamma - 2K > 0$, that is $K < \gamma/2$. Moreover, if $\gamma' > 0$, the condition

$$b = \gamma_n + \frac{(g + s\gamma')r^{*2}}{1 + sr^{*2}} > 0 \tag{9.14}$$

is automatically satisfied. Therefore, if the coupling strength $0 < K < \gamma/2$, the system (9.10) is asymptotically stable. Using Lyapunov indirect (linearization) method, the nonlinear system (9.9) is asymptotically stable around (r^*, N^*). That is, $\lim_{t\to\infty} r = r^*$ and $\lim_{t\to\infty} N = N^*$.

Now let's consider the dynamic of θ in Equation (9.5). As the right-hand side of this equation is continuous with respect to r and N, we have

$$\lim_{r\to r^*, N\to N^*} \dot{\theta} = \frac{\alpha}{2}\left(\frac{g(N^* - N_0)}{1 + sr^{*2}} - \gamma\right) + \omega$$

$$= -\alpha K + \omega \tag{9.15}$$

Therefore, $\lim_{t\to\infty}\dot{\theta} = -\alpha K + \omega$. This completes the proof.

9.4 Synchronization of Coupled Semiconductor Lasers

In this section, we consider n coupled semiconductor lasers described in (9.3). We can see that the system is a high-dimensional coupled nonlinear system. From the decoupled laser stability analysis in Section 9.3, we know that the limit cycle with a constant angular velocity is the dynamics of coupled lasers when they're fully synchronized. As we are interested in the local synchronization behavior around this equilibrium set, we can linearize the system around the limit cycle with $(r_k, N_k) = (r^*, N^*)$. Note that in synchronization, $\theta_k \to \theta_j, \forall k, j$.
  Define new states as follows:

$$\Delta r_k = r_k - r^*$$
$$\Delta N_k = N_k - N^*$$
$$\Delta\theta'_k = \theta_k - \bar{\theta} = \theta_k - \frac{1}{n}\sum_{i=1}^{n}\theta_i, \tag{9.16}$$

where r^*, N^* are defined in (9.7), $\Delta\theta'_k$ measures the difference between the phase of the kth laser and the mean phase of all lasers $\bar{\theta}$. We assume initially the phase differences between lasers are small. Linearizing the system (9.3) around the equilibria $(r_k, N_k, \Delta\theta'_k) = (r^*, N^*, 0), \forall k$, simplifying using

$$\sum_{i=1}^{n}\sum_{j\in\mathcal{N}(i)}(\Delta\theta'_j - \Delta\theta'_i) = 0$$

due to the symmetry of the coupling topology, the system equations of the error states can be written as

$$\Delta\dot{r}_k = A_{11}\Delta r_k + A_{12}\Delta N_k + \frac{K}{m}\sum_{j\in\mathcal{N}(k)}(\Delta r_j - \Delta r_k)$$

$$\Delta\dot{\theta}'_k = -B_1\left(\Delta r_k - \frac{1}{n}\sum_{i=1}^{n}\Delta r_i\right)$$

$$+ B_2\left(\Delta N_k - \frac{1}{n}\sum_{i=1}^{n}\Delta N_i\right) + \frac{K}{m}\sum_{j\in\mathcal{N}(k)}(\Delta\theta'_j - \Delta\theta'_k)$$

$$\Delta\dot{N}_k = A_{21}\Delta r_k + A_{22}\Delta N_k, \tag{9.17}$$

where $B_1 = \frac{a s \gamma' r^*}{1+s r^{*2}}$, $B_2 = \frac{a g}{2(1+s r^{*2})}$, A_{11}, A_{12}, A_{21}, and A_{22} are the scalar elements of matrix A defined in (9.11).

We can see that the equilibrium of the system (9.17) indicates synchronization, that is, at the equilibrium point $(\Delta r_k, \Delta \theta'_k, \Delta N_k) = (0, 0, 0)$, the states of the system (9.3) tend to $r_k \to r_j \to r^*, N_k \to N_j \to N^*, \theta_k \to \theta_j, \forall k, j$. Note that we linearize the coupled dynamics around $(r^*, N^*, \Delta \theta'_k) = (r^*, N^*, 0)$ as defined in (9.16), instead of following the convention of usual treatment by linearizing around (r^*, N^*, θ^*). This is because that the linearized model around (r^*, N^*, θ^*) always has a zero eigenvalue in its system matrix, which makes it difficult to analyze local stability. As shown later in this section, we draw local synchronization conclusions by the current treatment.

In matrix form, we have the following compact form:

$$\Delta \dot{r} = A_{11} \Delta r + A_{12} \Delta N - KL\Delta r$$

$$\Delta \dot{\theta}' = -B_1 L_0 \Delta r + B_2 L_0 \Delta N - KL\Delta \theta'$$

$$\Delta \dot{N} = A_{21} \Delta r + A_{22} \Delta N, \tag{9.18}$$

where $\Delta r = [\Delta r_1, \Delta r_2, ..., \Delta r_n]^T$, $\Delta N = [\Delta N_1, \Delta N_2, ..., \Delta N_n]^T$, $\Delta \theta' = [\Delta \theta'_1, \Delta \theta'_2, ..., \Delta \theta'_n]^T$, $L_0 = I - (\mathbf{1}\mathbf{1}^T/n)$, L is the Laplacian matrix defined in Assumption 9.1. For the case with all-to-all coupling, $L = L_0$.

We have Lemma 9.1 for the stability of (9.18).

Lemma 9.1   The linear coupled system (9.17) or its compact form (9.18) is asymptotically stable, if the system parameter $0 < K < \gamma/2$.

Proof: To prove asymptotical stability of (9.18), we first perform a similarity transformation. Define

$$z = \begin{bmatrix} z_1 \\ z_2 \\ z_3 \end{bmatrix} = M \begin{bmatrix} \Delta r \\ \Delta N \\ \Delta \theta' \end{bmatrix} \tag{9.19}$$

where

$$M = \begin{bmatrix} P & 0 & 0 \\ 0 & P & 0 \\ 0 & 0 & I \end{bmatrix}. \tag{9.20}$$

Here, P is the unitary matrix satisfying $PP^T = P^T P = I$ and $PLP^T = \Lambda_0$ with $\Lambda_0 = \text{diag}(\lambda_1, \lambda_2, ..., \lambda_n)$ being the eigenvalue matrix of L; that is, P is the orthogonal similarity transformation matrix transforming the symmetric matrix L into a diagonal one. With M defined in (9.20), Equation (9.19) can be rewritten as follows:

$$z_1 = P\Delta r \quad z_2 = P\Delta N \quad z_3 = \Delta \theta'. \tag{9.21}$$

Expressing the system (9.18) in the new coordinates yields

$$\dot{z}_1 = A_{11}z_1 + A_{12}z_2 - K\Lambda_0 z_1 \tag{9.22a}$$

$$\dot{z}_2 = A_{21}z_1 + A_{22}z_2 \tag{9.22b}$$

$$\dot{z}_3 = -B_1 L_0 P^T z_1 + B_2 L_0 P^T z_2 - KL z_3. \tag{9.22c}$$

Note that the given equations hold as $A_{11}, A_{12}, A_{21}, A_{22}$ are all scalers. Since $\sum_{k=1}^{n} \Delta\theta'_k = 0$ according to the definitions of $\Delta\theta'_k$ and $\bar{\theta}$, we have $\mathbf{1}^T z_3 = \mathbf{1}^T \Delta\theta' = 0$. Accordingly, $L_0 z_3 = (I - \frac{1}{n}\mathbf{1}\mathbf{1}^T)z_3 = z_3$. Also note that for $c_0 > 0$, we have $LL_0 = LL_0 + c_0\mathbf{1}\mathbf{1}^T L_0 = (L + c_0\mathbf{1}\mathbf{1}^T)L_0$ and $LL_0 = L(I - \frac{1}{n}\mathbf{1}\mathbf{1}^T) = L$. Comparing the right sides of these two equations, we have $L = (L + c_0\mathbf{1}\mathbf{1}^T)L_0$. Therefore, Equation (9.22c) can be rewritten as follows:

$$\dot{z}_3 = -B_1 L_0 P^T z_1 + B_2 L_0 P^T z_2 - K(L + c_0\mathbf{1}\mathbf{1}^T)L_0 z_3$$

$$= -B_1 L_0 P^T z_1 + B_2 L_0 P^T z_2 - K(L + c_0\mathbf{1}\mathbf{1}^T)z_3. \tag{9.23}$$

According to the spectral theorem [19], the eigenvalues of $L + c_0\mathbf{1}\mathbf{1}^T$ are c_0, $\lambda_2, \lambda_3, \dots \lambda_n$ with $\lambda_i > 0, i = 2, \dots, n$, denoting the ith smallest eigenvalue of L. Replacing (9.22c) with (9.23), we can rewrite the system dynamics of (9.22) as follows:

$$\dot{z}_1 = A_{11}z_1 + A_{12}z_2 - K\Lambda_0 z_1 \tag{9.24a}$$

$$\dot{z}_2 = A_{21}z_1 + A_{22}z_2 \tag{9.24b}$$

$$\dot{z}_3 = -B_1 L_0 P^T z_1 + B_2 L_0 P^T z_2 - K(L + c_0\mathbf{1}\mathbf{1}^T)z_3. \tag{9.24c}$$

The system matrix W of the above linear system is

$$W = \begin{bmatrix} A_{11}I - K\Lambda_0 & A_{12}I & 0 \\ A_{21}I & A_{22}I & 0 \\ -B_1 L_0 P^T & B_2 L_0 P^T & -K(L + c_0\mathbf{1}\mathbf{1}^T) \end{bmatrix} \tag{9.25}$$

To prove that z converges to zero, it is sufficient to prove that the eigenvalues of W locate on LHP. Note that W is a block lower triangular matrix, and the diagonal block $-K(L + c_0\mathbf{1}\mathbf{1}^T)$ has eigenvalues $-Kc_0, -K\lambda_2, -K\lambda_3, \dots, -K\lambda_n$, all of which locate on LHP. Therefore, we only need to prove that the following matrix

$$\begin{bmatrix} A_{11}I - K\Lambda_0 & A_{12}I \\ A_{21}I & A_{22}I \end{bmatrix} \tag{9.26}$$

has all eigenvalues on LHP in order to prove the fact that the eigenvalues of W locate on LHP. In fact, this is equivalent to the asymptotically stability of the following subsystem composed of z_1 and z_2:

$$\dot{z}_1 = A_{11}z_1 + A_{12}z_2 - K\Lambda_0 z_1$$

$$\dot{z}_2 = A_{21}z_1 + A_{22}z_2. \tag{9.27}$$

For the system (9.27), it is clear that this system is composed of n completely decouple subsystems. Therefore, its stability is equivalent to stability of all the subsystems. To show this more clearly, we write (9.27) into the following scalar form:

$$\dot{z}_{1i} = A_{11}z_{1i} + A_{12}z_{2i} - K\lambda_i z_{1i}$$

$$\dot{z}_{2i} = A_{21}z_{1i} + A_{22}z_{2i}. \tag{9.28}$$

Here, z_{1i} and z_{2i} are the ith component of vectors z_1 z_2, respectively. The system (9.27) is asymptotically stable if the following system matrices for all subsystems are Hurwitz:

$$A - K\lambda_i \begin{bmatrix} 1 & 0 \\ 0 & 0 \end{bmatrix} - - - - \text{Hurwitz.} \tag{9.29}$$

The characteristic polynomial for this matrix is as follows:

$$\lambda^2 + (\gamma_n + \frac{(g + s\gamma')r^{*2}}{1 + sr^{*2}} + K\lambda_i)\lambda$$

$$+ \frac{(g + \gamma_n s)\gamma' r^{*2}}{1 + sr^{*2}} + K\lambda_i(\gamma_n + \frac{gr^{*2}}{1 + sr^{*2}}) = 0. \tag{9.30}$$

Similar to the arguments in the decoupled laser case, the following inequalities are equivalent to the claim that the real parts of the roots to (9.30) are negative:

$$\gamma_n + \frac{(g + s\gamma')r^{*2}}{1 + sr^{*2}} + K\lambda_i > 0$$

$$\frac{(g + \gamma_n s)\gamma' r^{*2}}{1 + sr^{*2}} + K\lambda_i \left(\gamma_n + \frac{gr^{*2}}{1 + sr^{*2}}\right) > 0. \tag{9.31}$$

Recall that the eigenvalues of L satisfies $\lambda_1 = 0$, $\lambda_i \geq 0$ for $2 \leq i \leq n$. For $i = 1$, $\lambda_1 = 0$ and the dynamic (9.28) is identical to the decoupled laser case (9.10). For the inequalities in (9.31) to hold, we need $0 < K < \gamma/2$ as $\gamma' = \gamma - 2K$, with positive system parameters $g, s, \alpha, \gamma, \gamma_n$, and N_0, J. For $i = 2, 3, ..., n$, $\lambda_i > 0$, so (9.31) clearly holds. $\qquad\square$

Theorem 9.2 Under Assumption 9.1, if each laser is operating around the limit cycle with the radius r^*, the carrier number N^* and the angular velocity $(-\alpha K + \omega)$, and the initial phase differences of lasers are small (i.e., $\Delta\theta'_k$ is around 0), the coupled laser system (9.1) locally asymptotically synchronizes to the same limit cycle for the coupling strength $0 < K < \gamma/2$.

Proof: We showed at the beginning of this section that the system (9.1) and its polar coordinate representation (9.3) can be linearized to (9.17) around the limit cycle $(r_k, N_k, \Delta\theta'_k) = (r^*, N^*, 0), \forall k$, under the condition that the decoupled laser is stabilized around it and the initial phase differences are small. We can see that the equilibrium of the system (9.17) indicates synchronization of the system (9.3), that is, $r_k \to r_j \to r^*, N_k \to N_j \to N^*, \theta_k \to \theta_j, \forall k, j$. From Lemma 9.1, we know that (9.17) is asymptotically stable, which then indicates synchronization as $t \to \infty$. $\square$

Remark 9.3   From Theorem 9.2, the coupling topology to guarantee synchronization is a general undirected connected graph. When the lasers are synchronized, that is, $E_k = E_j$ for all possible k and j, the dynamic of each laser in the coupled array becomes identical with each other and is also identical to the behavior of the decoupled laser dynamics shown in (9.5).

9.5   Simulation Examples

To illustrate the performance, we consider five coupled semiconductor lasers modeled by (9.1). The system parameters used are listed in the Table 9.1, where the values are from a realistic experimental situation [18, 20].

Substitute these parameters into Equations (9.7) and (9.11), and choose $\omega = 2$, coupling strength $K = 1 < \gamma/2$. We have the periodic solution of each decoupled laser is $(r^*, N^*) = (302.2562, 1.8381 \times 10^8)$ with angular velocity $\dot{\theta}^* = 1.0001$, and

$$
A = \begin{bmatrix} -8.9361 & 0.0022 \\ -2.9565 \times 10^5 & -1.8458 \end{bmatrix}. \tag{9.32}
$$

The eigenvalues of A are $\lambda_1 = -5.3909 + 25.4089i$, $\lambda_2 = -5.3909 - 25.4089i$. According to Theorem 9.1, each decoupled laser system is stabilized on its periodic solution. In the presence of coupling, we simulate the case of all-to-all coupling and use the same parameters $\alpha, N_0, s, \gamma, \gamma_n$, and g and coupling strength K. According to Theorem 9.2, the coupled semiconductor

Table 9.1 Laser parameters that represent a realistic experiment.

Symbol	Description	Value
α	Linewidth enhancement factor	10^{-4}
γ	Photon decay rate	500ns^{-1}
γ_n	Carrier decay rate	0.5ns^{-1}
g	Differential gain coefficient	$1.5 \times 10^{-5}\text{ns}^{-1}$
N_0	Carrier numbers at transparency	1.5×10^8
s	Gain saturation coefficient	2×10^{-7}

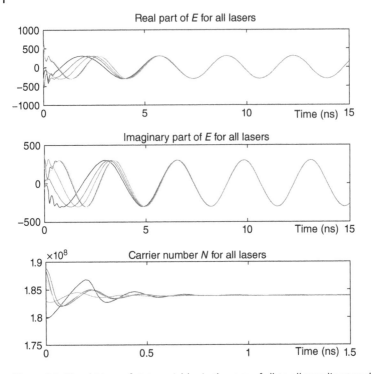

Figure 9.2 Time history of state variables in the case of all-to-all coupling topology.

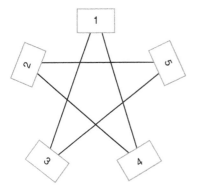

Figure 9.3 The coupling topology.

system will locally synchronize asymptotically. Figure 9.2 demonstrates the synchronization process of the system modeled by (9.17), which illustrates that $E_k(t) \rightarrow E_j(t)$, $\forall k, j = 1, \ldots, n$, as $t \rightarrow \infty$. In the case of general coupling topology, we choose the coupling for the five laser array as shown in Figure 9.3. Its synchronization process is shown in Figure 9.4. We can see that the convergence time is longer than the all-to-all coupling case.

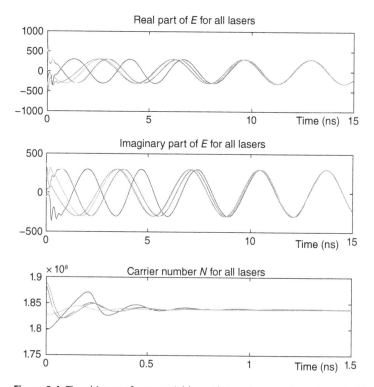

Figure 9.4 Time history of state variables with coupling topology shown in Figure 9.3.

9.6 Conclusion

In this chapter, we considered coupled semiconductor lasers modeled by nonlinear Lang and Kobayashi equations. We assume the coupling topology is modeled by an undirected connected graph with equal in-degrees. We first proved that the decoupled laser system is locally stabilized to a limit cycle under certain conditions on system parameters. Then, utilizing graph and systems theory, we rigorously proved that the coupled laser system locally synchronizes to the limit cycle. Simulations demonstrate the synchronization behaviors of five lasers in a generally connected topology.

Notes

The results in this chapter were first published in Ref. [21].

Acknowledgment is given to ©2013 IEEE. Reprinted, with permission, from Li, S., Guo, Y., and Braiman, Y. (2013) Synchronizing coupled semiconductor lasers under general coupling topologies, in American Control Conference, pp. 1231–1236.

References

1 Winful, H.G. and Wang, S.S. (1988) Stability of phase locking in coupled semiconductor laser arrays. *Applied Physics Letters*, **53**, 1894–1896.

2 Wang, S.S. and Winful, H.G. (1988) Dynamics of phase locked semiconductor laser arrays. *Applied Physics Letters*, **52**, 1774–1776.

3 Kozyreff, G., Vladimirov, A., and Mandel, P. (2000) Global coupling with time delay in an array of semiconductor lasers. *Physical Review Letters*, **85**, 3809–3812.

4 Kozyreff, G., Vladimirov, A., and Mandel, P. (2001) Dynamics of a semiconductor laser array with delayed global coupling. *Physical Review E*, **64**, 016 613.

5 Lang, R. and Kobayashi, K. (1980) External optical feedback effects on semiconductor injection laser properties. *IEEE Journal of Quantum Electronics*, **16** (3), 347–355.

6 Kuramoto, Y. (1984) *Chemical Oscillations, Waves and Turbulence*, Springer-Verlag, Berlin.

7 Botez, D. and Scifres, D.R. (1994) *Diode Laser Arrays*, Cambridge University Press, Cambridge, England.

8 Braiman, Y., Kennedy, T.A.B., Wiesenfeld, K., and Khibnik, A. (1995) Entrainment of solid-state laser arrays. *Physical Review A*, **52** (2), 1500–1506.

9 Fabiny, L., Colet, P., Roy, R., and Lenstra, D. (1993) Coherence and phase dynamics of spatially coupled solid-state lasers. *Physical Review A*, **47** (5), 4287–4296.

10 Garcia-Ojalvo, J., Casademont, J., Torrent, M.C., Mirasso, C.R., and Sancho, J.M. (1999) Coherence and synchronization in diode-laser arrays with delayed global coupling. *International Journal of Bifurcation and Chaos*, **9** (11), 2225–2229.

11 Hohl, A., Gavrielides, A., Erneux, T., and Kovanis, V. (1997) Localized synchronization in two nonidentical coupled lasers. *International Journal of Bifurcation and Chaos*, **23** (25), 4745–4748.

12 Hohl, A., Gavrielides, A., Erneux, T., and Kovanis, V. (1999) Quasiperiodic synchronization for two delay-coupled semiconductor lasers. *Physical Review A*, **59** (5), 3941–3949.

13 Winfree, A.T. (1980) *Geometry of Biological Time*, Springer-Verlag, New York.

14 Ruiz-Oliveras, F., Soriano, M.C., Colet, P., and Mirasso, C.R. (2009) Information encoding and decoding using unidirectionally coupled chaotic semiconductor lasers subject to filtered optical feedback. *IEEE Journal of Quantum Electronics*, **45**, 962–968.

15 Ermakov, I., Tronciu, V.Z., Colet, P., and Mirasso, C. (2009) Controlling the unstable emission of a semiconductor laser subject to conventional optical feedback with a filtered feedback branch. *Optics Express*, **17**, 8749–8755.

16 Jacobo, A., Soriano, M.C., Mirasso, C.R., and Colet, P. (2010) Chaos-based optical communications: Encryption vs. nonlinear filtering. *IEEE Journal of Quantum Electronics*, **46** (4), 499–505.

17 Oliva, R.A. and Strogatz, S.H. (2001) Dynamics of a large array of globally coupled lasers with distributed frequencies. *International Journal of Bifurcation and Chaos*, **11** (9), 2359–2374.

18 Liu, B., Liu, Y., and Braiman, Y. (2008) Coherent addition of high power laser diode array with a v-shape external talbot cavity. *Optics Express*, **16** (25), 20 935–20 942.

19 Meyer, C.D. (ed.) (2000) *Matrix Analysis and Applied Linear Algebra*, Society for Industrial and Applied Mathematics, Philadelphia, PA.

20 Liu, B., Braiman, Y., Nair, N., Lu, Y., Guo, Y., Colet, P., and Wardlaw, M. (2014) Nonlinear dynamics and synchronization of an array of single mode laser diodes in external cavity subject to current modulation. *Optics Communications*, **324**, 301–310.

21 Li, S., Guo, Y., and Braiman, Y. (2013) Synchronizing coupled semiconductor lasers under general coupling topologies, in *American Control Conference*, pp. 1231–1236, doi:10.1109/ACC.2013.6580004.

Appendix A

Notation and Symbols

$\mathbb{R}$	Field of real numbers
$\mathbb{C}$	Field of complex numbers
$\mathbf{I}$	Identity matrix
$\text{Re}\{\lambda\}$	The real part of $\lambda \in \mathbb{C}$
$\mathbf{1}$	Vector with all entries equal to 1
$\lvert a \rvert$	Absolute value of $a \in \mathbb{C}$
$\lVert \mathbf{x} \rVert_1$	1-norm of vector $\mathbf{x}$, defined as follows:
	$\lVert \mathbf{x} \rVert_1 = \lvert x_1 \rvert + \lvert x_2 \rvert + \cdots + \lvert x_n \rvert$ for $\mathbf{x} = [x_1, x_2, \ldots, x_n]^T$
$\lVert \mathbf{x} \rVert_2$ (or $\lVert \mathbf{x} \rVert$)	Euclidean norm of vector $\mathbf{x}$, defined as follows:
	$\lVert \mathbf{x} \rVert_2 = \sqrt{x_1^2 + x_2^2 + \cdots + x_n^2}$ for $\mathbf{x} = [x_1, x_2, \ldots, x_n]^T$
$\lVert \mathbf{x} \rVert_\infty$	Infinity norm of vector $\mathbf{x}$, defined as follows:
	$\lVert \mathbf{x} \rVert_\infty = \max_i \lvert x_i \rvert$ for $\mathbf{x} = [x_1, x_2, \ldots, x_n]^T$
$\mathbf{A}^T$	Transpose of matrix $\mathbf{A}$
$\mathbf{A}^{-1}$	Inverse of matrix $\mathbf{A}$
$\lambda_{\min}(\mathbf{A})$	Minimum eigenvalue of matrix $\mathbf{A}$
$\lambda_{\max}(\mathbf{A})$	Maximum eigenvalue of matrix $\mathbf{A}$
$\text{eig}(\mathbf{A})$	Eigenvalue of matrix $\mathbf{A}$
$\det(\mathbf{A})$	Determinant of matrix $\mathbf{A}$
$\text{trace}(\mathbf{A})$	Trace of matrix $\mathbf{A}$
$\text{vec}(\mathbf{A})$	Vectorization of matrix $\mathbf{A}$
$\text{diag}([\mathbf{A_1} \ldots \mathbf{A_n}])$	Block diagonal matrix with $\mathbf{A}_i$ as its ith element, $i = 1, \ldots, n$
$\mathbf{A} < 0$	Matrix $\mathbf{A}$ is negative definite
$\mathbf{A} \geq 0$	Matrix $\mathbf{A}$ is positive semidefinite
$\mathcal{N}(\mu, \sigma^2)$	Gaussian distribution with mean μ and variance σ^2
$\text{E}(\cdot)$	Expectation of a random variable
$\text{Var}(\cdot)$	Variance of a random variable
$\text{Pr}(\cdot)$	Probability of a random variable
$\text{span}(\cdot)$	The space spanned by a vector
$\text{argmax}(y(x))$	The arguments of the maxima: the point of x where $y(x)$ reaches its maximum
$\otimes$	Kronecker product, see Appendix B

Distributed Cooperative Control: Emerging Applications, First Edition. Yi Guo.
© 2017 John Wiley & Sons, Inc. Published 2017 by John Wiley & Sons, Inc.
Companion website: www.wiley.com/go/Guo/DistributedCooperativeControl

Appendix B

Kronecker Product and Properties

The Kronecker product is denoted by $\otimes$.

$$A \otimes B = \begin{bmatrix} a_{11}B & a_{12}B & \cdots & a_{1p}B \\ \vdots & \vdots & \vdots & \vdots \\ a_{n1}B & a_{n2}B & \cdots & a_{np}B \end{bmatrix}$$

Here, A is an $n \times p$ matrix and B is an $m \times q$ matrix. Some useful properties of Kronecker product are given in the follows:

1.

$$(A \otimes B)(C \otimes D) = AB \otimes CD$$

2.

$$A \otimes B + A \otimes C = A \otimes (B + C)$$

3.

$$(A \otimes B)^T = A^T \otimes B^T$$

4.

$$(A \otimes B)^{-1} = A^{-1} \otimes B^{-1}$$

5. The eigenvalues of $C \otimes D$ ($C \in \mathfrak{R}^{m \times m}, D \in \mathfrak{R}^{n \times n}$) are $\lambda_i \mu_j, i = 1, \ldots, m, j = 1, \ldots, n$, where $\lambda_i, i = 1, \ldots, m$ are eigenvalues of C and $\mu_j, j = 1, \ldots, n$ are eigenvalues of D.

Distributed Cooperative Control: Emerging Applications, First Edition. Yi Guo.
© 2017 John Wiley & Sons, Inc. Published 2017 by John Wiley & Sons, Inc.
Companion website: www.wiley.com/go/Guo/DistributedCooperativeControl

Appendix C

Quantization Schemes

Quantization schemes are used in Chapter 3. There are two types of quantization schemes utilized in existing literatures: deterministic [1, 2] and probabilistic [3]. Suppose the range and bits for quantization are $[-\eta, \eta]$ and b, respectively. Thus, we get $2^b - 1$ uniform intervals with a step of $\delta = 2\eta/(2^b - 1)$. The deterministic quantization rounds a real number x to $Q(x)$ according to the following:

$$Q(x) = -\eta + i\delta$$
$$\text{if } x \in [-\eta + (i - 0.5)\delta, -\eta + (i + 0.5)\delta),$$

Here, i is an integer. The probabilistic quantization rounds a real number $x \in [-\eta + i\delta, -\eta + (i + 1)\delta]$ in a probabilistic manner as follows [3]:

$$\begin{cases} P(Q(x) = -\eta + i\delta) = 1 - r \\ P(Q(x) = -\eta + (i + 1)\delta) = r. \end{cases}$$

Here, $r = (x + \eta - i\delta)/\delta$. As argued in Ref. [3], the quantization error introduced by the probabilistic quantization scheme is a random variable with zero mean and a finite variance.

References

1 Frasca, P., Carli, R., Fagnani, F., and Zampieri, S. (2009) Average consensus on networks with quantized communication. *International Journal of Robust and Nonlinear Control*, **19** (16), 1787–1816.
2 Nedic, A., Olshevsky, A., Ozdaglar, A., and Tsitsiklis, J. (2009) On distributed averaging algorithms and quantization effects. *IEEE Transactions on Automatic Control*, **54** (11), 2506–2517.
3 Fang, J. and Li, H. (2010) Distributed consensus with quantized data via sequence averaging. *IEEE Transactions on Signal Processing*, **58** (2), 944–948.

Distributed Cooperative Control: Emerging Applications, First Edition. Yi Guo.
© 2017 John Wiley & Sons, Inc. Published 2017 by John Wiley & Sons, Inc.
Companion website: www.wiley.com/go/Guo/DistributedCooperativeControl

Appendix D

Finite L_2 Gain

Chapter 3 uses the concept of L_2 gain. For a discrete-time sequence $x = (x_0, x_1, x_2, ...)$, its L_2 norm is defined as follows:

$$\|x\|_{L_2} \triangleq \left(\sum_{i=0}^{\infty} |x_i|^2 \right)^{\frac{1}{2}}. \tag{D.1}$$

The space $L_2[0, +\infty)$ is defined as the set of all discrete-time sequences x such that $\|x\|_{L_2} < +\infty$. From a system point of view [1], if we think $x \in L_2[0, +\infty)$ as a "well-behaved" input, the question to ask is whether the output y will be "well-behaved" in the sense that $y \in L_2[0, +\infty)$. To deal with unbounded "ever-growing" signals, an extended space $L_{2e}[0, \tau)$ is defined as follows:

$$L_{2e} = \{x | x_\tau \in L_2, \forall \tau \in [0, \infty)\}. \tag{D.2}$$

Here, x_τ is a truncation of the sequence x defined by

$$x_\tau(t) = \begin{cases} x(t) & 0 \le t \le \tau \\ 0 & t > \tau. \end{cases}$$

The extended space L_{2e} is a linear space that contains the unextended space L_2 as a subset. A mapping $H : L_{2e}[0, \tau) \to L_{2e}[0, \tau)$ is said to have a finite L_2 gain if there exists finite constants γ and β such that

$$\| (H(x))_\tau \|_{L_2} \le \gamma \|x_\tau\|_{L_2} + \beta \tag{D.3}$$

for all $x \in L_{2e}[0, \tau)$ and $\tau \in [0, +\infty)$ [1, 2].

References

1 Khalil, H. (2002) *Nonlinear Systems*, 3rd edn., Prentice Hall, Upper Saddle River, NJ.
2 Van der Schaft, A. (2000) *L2-Gain and Passivity Techniques in Nonlinear Control*, 2nd edn., Springer-Verlag, London.

Distributed Cooperative Control: Emerging Applications, First Edition. Yi Guo.
© 2017 John Wiley & Sons, Inc. Published 2017 by John Wiley & Sons, Inc.
Companion website: www.wiley.com/go/Guo/DistributedCooperativeControl

Appendix E

Radio Signal Propagation Model

Radio signal propagation model is used in Chapter 5. We represent the radio energy distribution in regular cells with each cell having a signal strength value. The radio signal strength of the ath cell with position s at time k is described as

$$y_a(k, s) = P_0(k) - L_a(k) - S_a(k, s) - F_a(k, s) - \iota_a(k), \tag{E.1}$$

where

1. $P_0(k)$ is the transmission power of the radio frequency (RF) transmitters in dB;
2. $L_a(k)$ is the free space propagation loss;
3. $S_a(k, s)$ is the median scale shadow fading loss;
4. $F_a(k, s)$ is the small scale fading, for example, multi-path or interference;
5. $\iota_a(k)$ is the environmental noise.

Specifically, assuming the transmitter trajectories are known, the free-space propagation loss, $L_a(k)$, is commonly modeled as

$$L_a(k)[dB] = L(d_0) + 10\Gamma \log(d_a(k)/d_0), \tag{E.2}$$

where d_0 is the reference distance and $d_a(k)$ is the distance between the RF transmitter and the ath cell. The small-scale fading, $F_a(k, s)$, is usually modeled as lognormal distribution and can be neglected for simplicity. The shadow fading, $S_a(k, s)$, for a propagation link from the RF transmitter to the ath cell can be modeled as [1–3] follows:

$$S_a(k, s) = \sum_{b=1}^{n_c} \omega_{ab}(k)x_b(s) + \epsilon_a(k, s). \tag{E.3}$$

Here, $x_b(s)$, $b = 1, ..., n_c$, is the attenuation occurring in cell b of the propagation link at time k, which is predetermined by the environmental geometry (assuming the environment geometry is temporally fixed); n_c is the number of

Distributed Cooperative Control: Emerging Applications, First Edition. Yi Guo.
© 2017 John Wiley & Sons, Inc. Published 2017 by John Wiley & Sons, Inc.
Companion website: www.wiley.com/go/Guo/DistributedCooperativeControl

cells covered by the propagation link; $\epsilon_a(k, s)$ is the model error; $\omega_{ab}(k)$ is the weighting ratio of the cell b, defined as

$$\omega_{ab}(k) = \begin{cases} \frac{1}{\sqrt{d_a(k)}} & \text{if } \kappa_b + \chi_b < d + \lambda \\ 0 & \text{otherwise} \end{cases} \tag{E.4}$$

where κ_b is the distance between the cell b to the RF transmitter, χ_b is the distance between the cell b and the cell a, and λ is the width of the propagation link.

References

1 Patwari, N. and Agrawal, P. (2008) Effects of correlated shadowing: Connectivity, localization, and rf tomography, in *Proceedings of International Conference on Information Processing in Sensor Networks*, pp. 82–93.

2 Agrawal, P. and Patwar, N. (2009) Correlated link shadow fading in multi-hop wireless networks. *IEEE Transactions on Wireless Communications*, **8** (8), 4024–4036.

3 Wilson, J. and Patwari, N. (2010) Radio tomographic imaging with wireless networks. *IEEE Transactions on Mobile Computing*, **9** (5), 621–632.

Index

Distributed Cooperative Control: Emerging Applications, First Edition. Yi Guo.
© 2017 John Wiley & Sons, Inc. Published 2017 by John Wiley & Sons, Inc.
Companion website: www.wiley.com/go/Guo/DistributedCooperativeControl